KFS 对话建筑
Architecture Forum

KFS DESIGN IN 15 YEARS
KFS 设 计 15 年

INTERNATIONAL ARCHITECTURAL SERIES

KFS DESIGN INTERNATIONAL INC.

SELECTED AND CURRENT WORKS

KFS 国际建筑师事务所作品精选

2001—2015

KFS 国际建筑师事务所 编

大连理工大学出版社

图书在版编目(CIP)数据

KFS对话建筑：KFS设计15年/KFS国际建筑师事务所编.—大连：大连理工大学出版社,2015.10
ISBN 978-7-5685-0135-4

Ⅰ.①K… Ⅱ.①K… Ⅲ.①建筑设计—世界—现代—图集 Ⅳ.①TU206

中国版本图书馆CIP数据核字（2015）第229193号

出版发行：大连理工大学出版社
（地址：大连市软件园路80号 邮编：116023）
印　　刷：利丰雅高印刷（深圳）有限公司
幅面尺寸：220mm×300mm
印　　张：18
插　　页：4
出版时间：2015年10月第1版
印刷时间：2015年10月第1次印刷
责任编辑：裘美倩
责任校对：林　影
版式设计：倪文华　周康林　陆晓峰
封面设计：倪文华

ISBN 978-7-5685-0135-4
定　　价：258.00元

电 话：0411-84708842
传 真：0411-84701466
邮 购：0411-84708943
E-mail：designbookdutp@gmail.com
URL：http://www.dutp.cn

如有质量问题请联系出版中心：（0411）84709043　84709246

设计15年，KFS设计在中国

从2001年至今，KFS走过了漫漫15年的路。

城市与建筑是人类共同拥有的。从2005年1月1日起KFS设计以其特有的社会责任感，首次在上海教育电视台投资创办了一档周播的建筑普及及专业访谈节目——每周六晚9点40分的《KFS对话建筑》。节目和观众相约建筑，共同参与建筑与城市的发展。

《KFS对话建筑》受到了社会各界的广泛关注和赞誉。KFS也赢得了社会的良好赞誉。KFS踏踏实实地走过了15年，坚持建筑师的社会责任心，力求对社会、对城市有所贡献，使城市和建筑不留下遗憾和瑕疵。

KFS努力将先进的设计理念与中国的区域文化结合起来，将建筑师的责任与社会的需求结合起来，创造了如"上海滩花园洋房"等一系列令人称道的好作品。同时也在努力涉足城市建筑房地产领域，例如，"90花园别墅"。

"设计15年——设计创造价值"是KFS一直秉承的宗旨。强调建筑的独创性，通过独特的规划建筑构思，为城市和社会带来更多的价值。

整整15年过去了，回首看KFS的15年的步伐，仍是那样稳健和从容，KFS力求继续走在行业的前列，要求自己不断地洞察行业形势、引领多元发展。尽管很辛苦，但也很快乐。

未来的KFS希望带给城市与建筑的理念是：社会、城市、建筑走向多元和复合，充满人性，充满活力。

傅国华
博士. 总裁

2015.9

Design in 15 years, KFS Design in China

Since 2001, KFS has went through a 15-year architectural journey.

The city and architecture belongs to the people. Quality design significance was brought to the public via the K.F.Stone television series entitled "KFS. Architecture Forum", 9:40 p.m. every Saturday on the Shanghai Education Channel since 1st January 2005. International design ideologies and views of government officials in art and architecture were presented with interviews and features of design awards in architecture.

"KFS Architecture Forum" received positive feedback and interest from different sectors of society since its inception as did KFS design work. KFS has made a 15-year-long journey with the fundamental responsibility of architects to society, with contributions to both the city and it's people without compromise.

KFS has strived to unite modern design ideas with Chinese local culture, harmonizing an architect's design values with public needs, resulting in projects such as Shanghai Bund Villas and other architectural works. KFS is also working in real estate projects such as the 90 Square-Meter Villas.

"KFS Design in 15 years - Design Creates Value" is and has always been the pre-eminent principle of KFS. KFS creates uniqueness for the architecture through authentic concepts of urban planning, providing increased value for the city and society.

15 years has gone by, and through all these time, KFS maintains its design quality and its position in the architectural field, and is still expanding and upgrading itself. It demands devotion, but results in joy.

In the future, KFS has the hopes to provide diversity for the society and the city to fulfil the needs of the urban citizen.

Peter Fu Ph.D.
President.
Sept. 2015

加拿大总理史蒂芬·哈珀与傅国华博士，2009
Canadian Prime Minister Stephen Harper and Dr.Peter Fu, 2009

前加拿大总理让·克雷蒂安与傅国华博士，2004
Former Canadian Prime Minister Jean Cretien and Dr.Peter Fu, 2004

目录		CONTENTS
城市设计		**URBAN DESIGN**
2010年上海世界博览会申博方案之一	10	Shanghai World Expo. 2010 Bid Proposal Submission
上海"一城九镇"之北美风情金山枫泾规划	12	"North American Style Town" Planning, Fengjing, Shanghai
上海青浦城区中心区规划与实施	14	Qingpu Central Area Planning And Implementation, Shanghai
上海浦东锦绣华城规划与实施	18	Graceful Oasis City Planning And Implementation, Shanghai
俄罗斯圣彼得堡波罗的海明珠	30	Baltic Pearl, St.Petersburg, Russia
哈尔滨爱建滨江国际（爱建新城）规划与实施	34	Aijian New City Planning And Implementation, Harbin
海南三亚亚龙湾一号规划设计	50	1# Yalong Bay Planing, Sanya, Hainan
长泰锡东国际社区	58	Changtai Xidong International Community, Wuxi
西昌邛海凤凰谷度假小镇	60	Phoenix Resort Valley, Qionghai, Xichang
常熟虞山尚湖水中花	66	Yu Mountain Shang Lake Eco City, Changshu
公共建筑		**COMMERCIAL DESIGN**
上海海上文博苑方案设计	70	Haishang Cultural Museum Complex, Shanghai
国际乒乓球联合会博物馆设计方案	78	Proposal for International Table Tennis Museum, Shanghai
珠海横琴岛横琴大酒店	80	Hengqin Hotel, Hengqin, Zhuhai
上海浦东锦绣华城假日酒店	84	Holiday Inn Pudong, Shanghai
无锡千禧大酒店	90	Millennium Hotel, Wuxi
成都新东方千禧大酒店	96	Millennium Hotel, Chengdu
珠海横琴岛中大金融大厦	100	Zhongda Financial Tower, Hengqin, Zhuhai
上海浦东张江长泰国际广场	108	Changtai International Plaza, Shanghai
海南三亚亚龙湾一号商业综合体	112	1# Yalong Bay Commercial Building, Sanya, Hainan
珠海淇澳游艇会	118	Qi'ao Yacht Club, Zhuhai
哈尔滨爱建滨江国际交银大厦	122	Communication Bank Tower, Harbin
上海长宁舜元大厦（北大青鸟）	126	Shunyuan Office Tower (Beida Qingniao), Shanghai
上海浦东世纪大道长泰国际金融大厦	128	Changtai Office Tower, Pudong, Shanghai
上海浦东东晶国际办公大厦	132	Dongjing International Office Complex, Pudong, Shanghai
哈尔滨爱建滨江国际SOHO(居住办公一体建筑)	136	Aijian SOHO, Harbin
上海加拿大梦加园	140	Dream Home Canada, Shanghai
上海海纳科技研发大楼	144	Haina Hi-Tech Building, Shanghai
派诺珠海科技园	148	Pilot High-Tech Park, Zhuhai
KFS国际建筑师事务所上海办公楼	150	KFS Office Building, Shanghai
上海苏河一号（华森钻石广场）	156	1# Suzhou Creek (Huasen Diamond Plaza), Shanghai
上海静安达安河畔雅苑	160	Da'an Riverside Tower, Shanghai
珠海金山软件园区	164	Kingsoft Headquarters, Zhuhai
杭州西湖勾山国际	172	Goushan International, the West Lake, Hangzhou
苏州唯亭君地一期	178	Weiting Phase 1, Suzhou

居住建筑		**RESIDENTIAL DESIGN**
三亚亚龙湾一号居住综合体	184	1# Yalong Bay, Residential Building, Sanya
上海松江安贝尔花园	188	The Albert Garden, Songjiang, Shanghai
南京秦淮河G65项目	192	G65 Qinhuai River Project, Nanjing
青岛惜福镇泰晤士小镇	196	Thames Town, Xifu, Qingdao
无锡日式服务公寓	200	Japanese Style Service Apartment, Wuxi
上海古北国际花园	206	Gubei International Garden, Shanghai
上海浦东上海滩花园洋房	210	Shanghaitan Garden House, Shanghai
上海达安圣芭芭花园河谷3号—90花园别墅	214	Da'an St.Babara Valley No.3 - 90Villa, Shanghai
上海达安崇明御廷—90花园别墅	220	Da'an Royal Garden - 90Villa, Shanghai
上海绿地南汇布鲁斯小镇—90花园别墅	222	Bruce Town - 90Villa, Shanghai
上海静安达安花园(部分一期除外)	224	Da'an Garden, Shanghai (Excludes part of phase 1)
上海静安达安锦园	226	Da'an Jin Garden, Shanghai
上海普陀愉景华庭	238	Yujing Garden, Shanghai
上海黄浦明日星城(部分一期除外)	232	Tomorrow Star City, Shanghai (Excludes part of phase 1)
上海浦东东晶国际公寓	234	Dongjing International Residential, Shanghai
上海长宁春天花园	238	Spring Garden, Shanghai
上海虹口明佳花园	240	Mingjia Garden, Shanghai
上海徐汇电影华苑	242	Cinema Garden, Shanghai
成都上海花园	244	Shanghai Garden, Chengdu
上海绿地成都维多利亚花园	246	Victoria Garden, Chengdu
杭州瑞城花园·格林兰登度假庄园	248	Ruicheng Greenland Island, Hangzhou
上海徐汇漕河景苑	252	Caohejing Garden, Shanghai
上海宝山旭辉依云湾	254	LA BAIE D`EVIAN, Shanghai
上海徐汇百汇苑二期	258	Baihui Garden Phase 2, Shanghai
哈尔滨松江新城	260	Songjiang New city, Harebin
青岛南山果岭艺墅	264	Nanshan Golf Villas, Qingdao
室内设计		**INTERIOR DESIGN**
无锡千禧大酒店室内设计	268	Interior Design of the Millennium Hotel, Wuxi
上海达安圣芭芭花园河谷3号—90花园别墅室内设计	278	Interior Design of St.Babara Valley No.3-90 Villa, Shanghai
KFS国际建筑师事务所上海办公楼室内设计	284	Interior Design of KFS Office, Shanghai

URBAN DESIGN

城市设计

2010年上海世界博览会申博方案之一
Shanghai World Expo 2010 Bid Proposal Submission

中国，上海 Shanghai, China
业主：上海市城市规划管理局
Client: Shanghai Urban Planning Administration Bureau
设计时间 Design: 2001

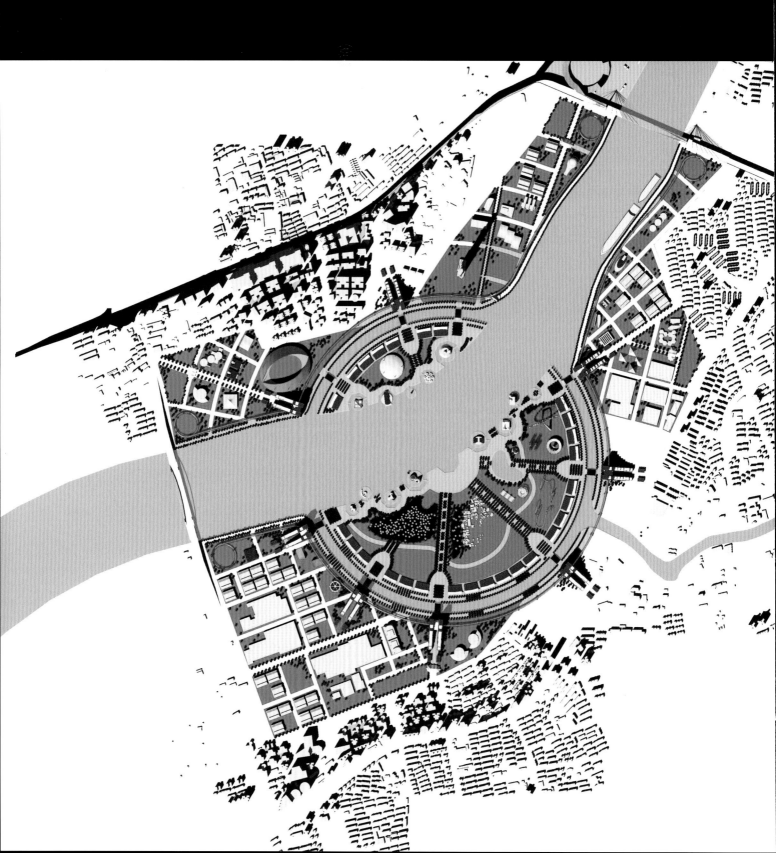

KFS非常有幸成为上海市政府所邀请的七个申博征集方案之一的设计单位——唯一的一个来自北美的设计单位。

2010年上海世博会的申博方案设计灵感来源于现存的地理环境特征：依现有黄浦江支流的走势，加部分人工河道，将展会的主体内容沿着这条圆形的轴线来布置。轴线的中央是大型绿地和休闲设施。整个规划区域由于被黄浦江分隔，因此如何将人流运输到对岸也成为本次设计主要考虑的问题之一。

世博会园区由永久住宅、综合建筑、多国综合展馆、独立国家展馆、主题公园及绿地等组成。设计的概念是：设计绿色岛屿，以减少基地被黄浦江明显分隔的形态而使之成为一个整体，同时强调一个主题——围绕水体来创造一个更美的城市。

北部扇形区有其他永久性的展馆用地和一个巨大的开放区域。在人工湖半径之外，黄浦江南侧是留给主要国家作展馆的用地。圆是整个概念的主题，八个塔式建筑将分布于圆周围，成为限定圆形空间的"主柱"，它们像一面虚设的墙围合了在世博会后将建造的21世纪城市中心的一个美丽的花园。作为2010年世博会主题的一部分，设计提倡对可二次利用和可循环使用的材料的开发利用。

世博会共开展180天，总参观人次预计达7 000万人。为适应高峰时平均每天约70万人次的交通量，该地区的建筑容积率可达4.5左右。设计面临的挑战是：需寻找一个能在世博会结束后同样对该地区有长期积极作用的规划。世博会的后续使用中，作为整个平面耀眼之处的圆形岛，在世博会结束之后将成为一座公园被保留下来，其包括了完整的主题公园、山丘、小道和人工河。发射状的林荫大道也将被保留下来，将来建造商业、娱乐、餐馆、购物中心等设施。一部分结构将成为户外开放空间。圆形人工河之外的塔式建筑将会留作酒店或变成居住建筑。以上都体现了"城市，让生活更美好"这一世博主题。

总用地面积：　　　　　　约500 ha.

The Government of Shanghai invited seven architecture firms in 2001 to participate in this limited design competition for the planning and design for the World Exposition 2010 site. KFS, as the only participant from North America was honored, and privileged to have received this prestigious invitation.

The inspiration of the planning concept was derived from the current geographic characteristics. Additional channels to the existing Huangpu River were provided. The main content of the exposition was placed in circular radiating lines. At the centre of this ring was a large public park open space. The Huangpu River divided the site into two parts. The main area of concern was the communication and integration of these two parts.

The exposition site contains permanent residential units, multi-functional buildings, national and multi-national pavilions, theme parks and green spaces. The KFS concept was the creation of a green island that reduced the division of the site by the river. This island simultaneously emphasized, the theme of a better city using water.

KFS located several permanent pavilions and a large public space at the northern part of the site. The southern part was reserved for national pavilions. Circulation was a theme applied to the site – eight "towers" are placed around the circular perimeter of the site. They formed a grid of pillars that defined the circular space of a future garden in the city of Shanghai. As part of the theme of this exposition, KFS advocated the development of re-used and re-cycled materials.

During the 180 days of exposition the projected estimate of the total visitor number would exceed 70 million bringing approximately 0.7 million visitors to the site every day. The F.A.R of the site was at least 4.5 and resulted in the many design challenges. A design solution with a permanent impact on the local development was mandatory. The site would be transformed into a major park at the conclusion of the 2010 Exposition. This included the theme park, the hills, the paths, and creeks. A radial network of boulevards was also preserved and to become the growing space for future commercial and leisure facilities. The tower buildings around the circular artificial river would be transformed into hotels or residential units. The ground level floors of the towers would continue as commercial uses. The design exemplifies the expo theme of "better city better life".

Site area:　　　　　　　500 ha.

上海"一城九镇"之北美风情金山枫泾规划
"North American Style Town" Planning, Fengjing, Shanghai

中国,上海 Shanghai, China
业主:上海金山区规划管理局
Client: Shanghai Jinshan Urban Planning Administration Bureau
设计时间 Design: 2003

北美风情镇的设计，来自棒球的突发灵感

设计就是构建一个生活环境，或者说是营造一种生活方式。金山区枫泾镇是上海的西南门户——江南名镇之一，已有1500多年的历史。作为上海"一城九镇"之一的枫泾镇，设计的目的就是设计一个体现北美悠闲、舒适生活节奏的小城镇，体现"生活方式高于建筑形式"的理念。

规划设想从建构北美特色风貌入手，镇区中心由中心风貌区、80 m宽的林荫大道及门户公园组成。东侧铁路沿线的门户公园是高速公路进入风情城的必经之路，由特色果园及购物公园组成，构成北美城镇的典型要素。门户公园既可减少铁路对城镇中心的影响，又是该地区与其他区域的连接点，成为北美特色风貌的一个外向型展示空间。

基地内设置的一所小教堂、垒球场以及会所，使北美人士亦能感受到家乡的文化气息和生活方式。高尔夫推杆果岭、红枫叶、摘草莓、打棒球、大片的葡萄种植地让酿酒成为可能，结合门户公园、大型超市等等，形成了北美生活的最重要的构成部分。

规划设想中心密、周边疏的布局模式，密度高的办公、金融及公共服务设施中心被设置于地块的西南中心湖区域。占地大、密度低的别墅区被布置于外围部分，围合式的多层住宅位于中部。

该方案已作为实施方案。

总用地范围：	约540 ha.
总建设用地面积：	约417 ha.
规划居住人口：	约2.8万人

The design intent provided a life-style in Fengjing county of Jinshan District of Shanghai. This historically important county played a significant new role in Shanghai's urban planning. The design of an ecological friendly and multi-functional town offered a relaxed North American inspired life-style. This philosophy is best illustrated in the words –"life style as an architectural expression".

The centre core recalled the character of a typical North American town. 80 m wide boulevard and parks dominate the centre of the town. Together with the shopping parks and the fruit gardens, they portrayed a classic image of North American life style. This big green zone shielded the town from the impact of the railway that connected the surrounding areas. A showroom of the North American life style was created.

Baseball diamonds, golf courts, maple leaf trees, berry picking, social clubs, and even a small church was provided in this community to bring the expatriates closer to their home culture, while large areas of grape planting made local brewing a reality.

The density of planning, in a radial manner in the core area to the west of the lake consisted of office buildings, commercial functions and living facilities. The villa area is on the outskirts of the town while the condominiums were located closer to the centre.

The planning and design were approved and the development was completed as designed.

Site area:	540 ha.
Land used:	417 ha.
Inhabitants planned:	28,000 people

上海青浦城区中心区规划与实施
Qingpu Central Area Planning And Implementation, Shanghai

中国，上海 Shanghai, China
业主：上海青浦城市规划管理局
Client: Qingpu Urban Planning Administration Bureau
设计时间 Design: 2001
建成时间 Completion: 2009

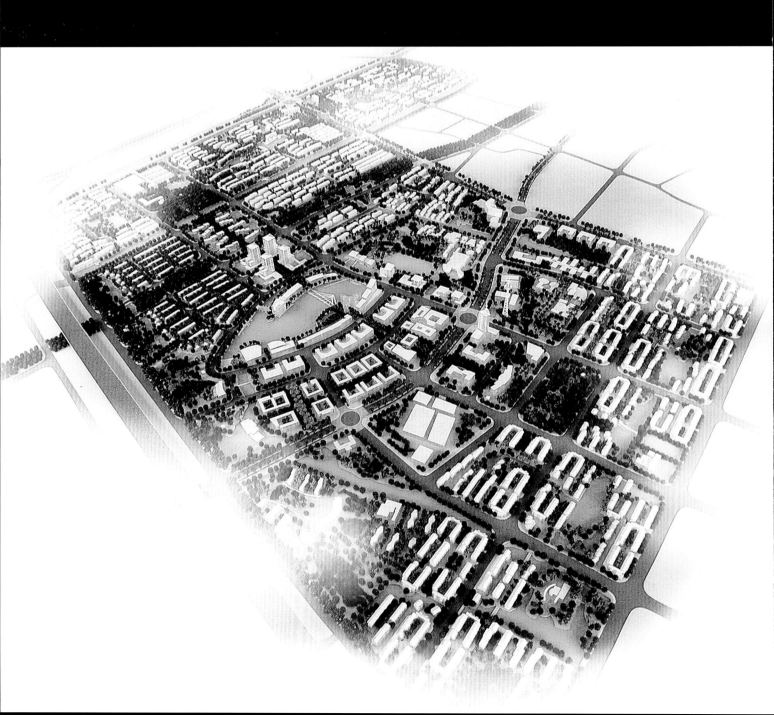

沪青平高速公路进入中心城区的入城口水景与建筑

青浦城区中心区的规划始于2001年初的一个国际竞赛，KFS在竞赛中拔得头筹，之后即进入实施阶段，2009年初具雏形。

青浦更新计划的主要构思之一是创建一个联结三个区域的城市绿化带。这三个区域分别为现有老城区与兴建中的新城区以及在新计划中的展览及体育运动区域。

青浦新建的绿化带将是一个重要的生态区，除了改善空气质量的功能外，绿化空间中还包含一些公共休闲娱乐设施。绿化通道在新老城区间形成了一定的分隔，使居民或游客对城市空间有更清晰的认识。

在绿化带中，设计了一系列富有特色的"城市生活空间"，它们展示了生动的城市形象和层次。沪青平高速公路的出口处设计了水景及富有当地特色的建筑物。在绿化带的东南位置，设置了一块扇状滨海公寓特别区。

设计中一个最大的亮点即是利用现有水面进行梳理和人工挖掘，形成青浦最具特色的"夏阳湖"。经过八年的建设，今天的青浦城区中心区面貌已焕然一新，"夏阳湖"的建成，给青浦增添了无限的色彩。夏阳湖畔已成为集青浦图书馆、博物馆、市民休闲、居住的文化活动中心。

The planning of the district of Qingpu was a competition that took place in 2001 with construction commencing shortly after KFS design submission was the successful entry. In 2009 the basic design and character emerged with praise and acceptance.

The primary design element was the upgrading of this district through the intergration of the green zones with the ones of neighboring districts – an existing old town, a new zone under construction and a sport and public zone were planned.

The new green belt of Qingpu served as an important ecological zone that helped to control the local air quality. It offered a number of public spaces inside itself and acted as a boundary that would give people a clearer definition of the surrounding space.

A series of "urban living spaces" was situated on the green belt to demonstrate the notion of local life and urban image. On the waterfront, south-west of the green belt, a prestigious group of condos was located. Waterscape and urban sculptures with local themes were designed close to the highway exit.

The lake in the centre of the site was the feature of the project. By modifying the existing water network, KFS created the "Xiayang Lake". It is now a significant urban part of the district centre as its banks offer pleasant public spaces and hosts Qingpu Library, museums and a culture centre.

规划总用地面积：	452 ha.
规划总建筑面积：	2 500 000 m²
规划居住总人数：	51 000 人

Site area:	452 ha.
Gross floor area:	2,500,000 m²
Inhabitants planned:	51,000 people

青浦夏阳湖

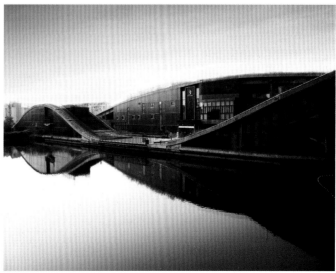

"锦绣华城"是一个从规划构思到建筑实施的典型成功案例。整个设计团队用了十年时间,完成了近2 000 000 m²的建筑量。今天已初具规模,成为上海浦东的最大居住区之一。

"锦绣华城"位于杨高路东、博文路南、锦绣路西、川杨河北,占地345 ha.,实际居住区用地313 ha.。拥有居住人口86 000人。区域内建筑种类较多,主要以居住建筑为主,还包含高档酒店——锦绣假日酒店、中小学校及各种类型的商业建筑等。

设计构思上,以"泛公园"的理念为主,把一个3.45 km²的居住区设计成一个以生态为主题的大公园,规划构筑了一个分布在3.45 km²内的绿色网格构架。这个"绿网"以基地内的高压走廊、河流、城市道路为主要骨架,再把每个小区的中心绿地连接上去,最后发散到每个组团绿地。这张"绿网"有机地与城市总体格局相呼应,与基地现状相适应。

成山路作为交通型的城市道路,设有多处公交站点和地铁出入口。利用成山路的交通和人气的优势,将酒店及办公等商业服务设施集中布置在沿路两侧。商业入口广场与居住绿地的互相结合,形成都市特有的居住区概念的商业模式。

该方案已作为实施方案。

用地面积:	345 ha.
规划建筑面积:	2 900 000 m²
规划居住人口:	86 000人

The Graceful Oasis City is a successful example of planning, and architectural design. Today it is one of the largest residential areas in Pudong with the completion of 2,000,000 m².

The site of Graceful Oasis project is defined by four urban elements: Yanggao Road to the East, Bowen Road to the South, Jinxiu Road to the West, and Chuanyang Creek to the North. The site area is 345 ha., with 313 ha. for residential use. The current population is 86,000. Among various types of architecture in the site the majority is residential, combined with elegant Jingxiu Holiday Inn, educational facilities and different commercial units.

"Pan-Park" is the principal design element in this project. Transforming a 3.45 km² residential zone into a large scale ecological-theme park. The plan was the creation of a green network within a 345 ha. plot by using high-voltage grids, rivers, and urban road as the fundamental structure. The green zone of each neighborhood was connected to this structure, forming the green web that existed in the urban texture.

Chengshan Road was the main traffic corridor through the site, with multiple bus stops and subway stations. To live up to this potential KFS placed hotels, offices and commercial functions along this road. The main commercial entrance merged with the green area of the residential zones, forming a model with a unique commercial character.

Site area:	345 ha.
Gross floor area:	2,900,000 m²
Inhabitants planned:	86,000 people

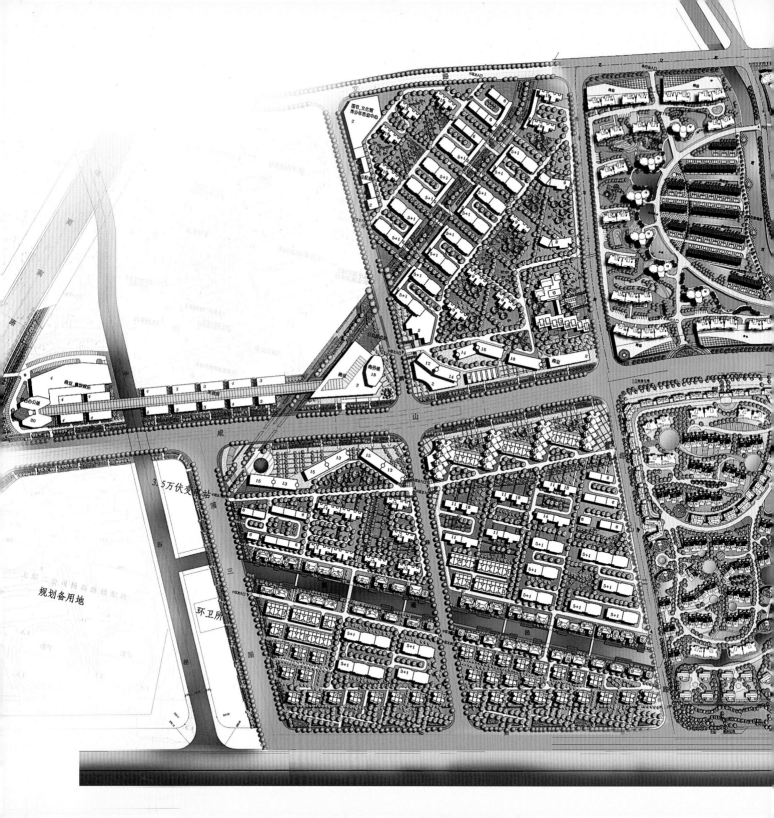

总平面图 Site Plan

俄罗斯圣彼得堡波罗的海明珠
Baltic Pearl, St.Petersburg, Russia

俄罗斯，圣彼得堡 St.Petersburg, Russia
业主：波罗的海明珠股份有限公司
Client: Baltic Pearl Co.,Ltd.
设计时间 Design: 2007

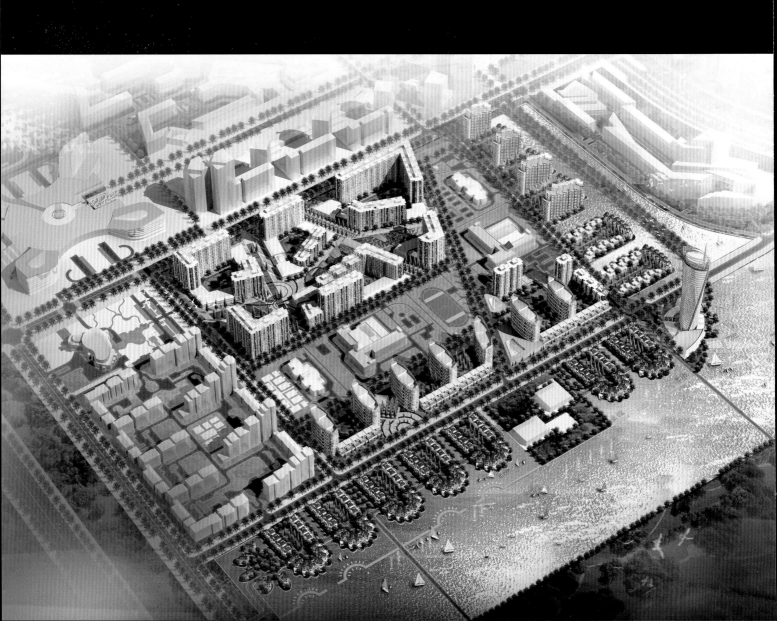

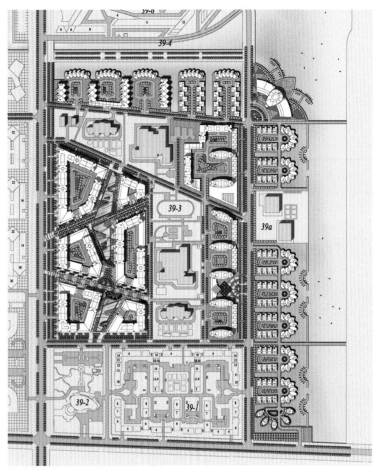

总平面图　Site Plan

"波罗的海明珠"整个区域占地208 ha., 本项目的设计范围是"波罗的海明珠"中的39-3地块和39a地块。

本项目的规划用地约29 ha., 地上总建筑使用面积约为447 000 m²。基地位于整个区域的东南侧,是以居住为主,结合商业、教育等设施的高档居住社区。基地东侧和北侧紧临杜杰尔戈夫斯基运河及其支流,东侧河宽逾百米,对岸即为列宁格勒公园,环境资源极为优越。

设计旨在营造传统城市形态,延续圣彼得堡的城市脉络,最大化拓展周边景观资源,提升住宅价值,通过规划增强社区意识。设计从每个地块的特殊性出发,以不同的设计态度及手段对其进行处理,彰显其独特个性。同时,设计还选择多样化的住宅种类,丰富区域的建筑形态特征,优化居住套型。

各个居住地块主要采用小尺度围合型居住组团,组团的外部成为开放的公共空间,组团内部有强烈的内聚感,通过与外部空间的适当隔离,形成了向心空间,强化了社区意识。沿河区域主要设置联排别墅,充分利用景观资源,做到户户有"水景"。

靠近基地西侧——特里布查路的地块,设计以周边地区将有较为密集的商业办公建筑,城市特征非常明显,故而设计以板式高层居住组团配以底层商业,促成丰富的城市生活。

总用地面积:	29 ha.
地上总建筑使用面积:	440 000 m²
规划居住人口:	13 000人

This project was part of the large-scale Baltic Pearl development, with a total site area of 208 hectares. KFS designed the plots 39-3 and 39a.

The two plots are located at the south-east of the site. They occupy 29 ha. and have a gross floor area of 447,000 m². With supporting commercial and educational facilities they are planned to be high-end residential communities. The site borders Dudergofsky Canal and branches into the north and east. Leningrad Park is at the other side of the river.

KFS took measures to have the traditional St. Petersburg urban texture grow into this site. Some of the goals aimed to maximize value of the surrounding land in order to increase the value of the residential units and the augmentation of community-awareness through sound planning and design. The designs were based on the individual characteristics of each plot. Different approaches were used to analyze and process different blocks in order to give them more personality. KFS chose various types of residential units to make the community richer in architectural language and vocabulary.

The majority of the residential units were small-scale closed clusters. They formed an outer public area and an inner private space. This increased the local community-awareness. Town houses are placed along the river to take advantage of the water view.

In the surrounding areas of Tributsa Road greater density commercial and office building were planned with a strong sense of urban environment. Together with wide highrise units they meet the functional needs of a modern urban lifestyle.

Site area:	29 ha.
Gross floor area:	440,000 m²
Inhabitants planned:	13,000 people

哈尔滨爱建滨江国际（爱建新城）规划与实施
Aijian New City Planning And Implementation, Harbin

中国，哈尔滨 Harbin, China
业主：哈尔滨爱达投资置业有限公司
Client: Harbin Aida Investment Real Estate Co.,Ltd.
设计时间 Design: 2002
建成时间 Completion: 2009

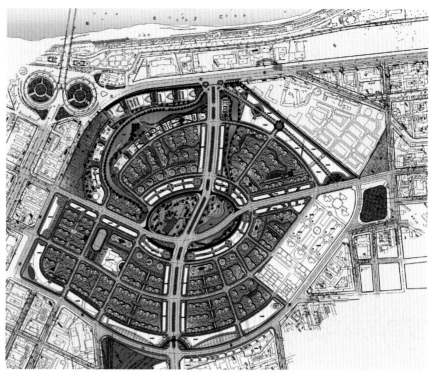

总平面图　Site Plan

哈尔滨爱建滨江国际社区（爱建新城）规划始于一个影响力很大的国际设计竞赛。KFS力挫群雄，赢得其规划并进行了近八年的规划实施，完成了近90%（约2 000 000 m²）的建筑设计。其中包括一系列的哈尔滨标志性建筑（如交银大厦等）和大量的居住建筑。

爱建滨江国际社区（爱建新城）位于哈尔滨旧城区松花江畔，原址是著名的哈尔滨车辆厂——中国最早、规模最大的火车制造厂。

设计将新概念引入新的城市设计理念，借鉴一些可以促进经济发展、提高文化层次、提升美学和历史价值的元素，同时尊重本地的城市脉络，延续并创造富有哈尔滨城市情节的空间环境，注重保护自然生态，营造一个生机勃勃、绿意盎然的生态新城。

新城在遵循原存的城市肌理的基础上利用原有的工厂大型广场，新开辟出椭圆形的大片公园（绿地），以开阔的绿地在较高密度的新城中创造出宜人的空间。周围的多层商业建筑围绕中心公园展开，围合出新城中心地带。其建筑造型主要运用现代建筑语言并借鉴当地历史和地域特色。环形商业区的外层是高层住宅，面向中心公园，为住户提供了良好的景观。

新城商业建筑成系统布置，并沿主要街道两侧布置。商业建筑以多层及低层为主，在内部形成亲切、宜人的城市街道空间。整个新城商业气氛浓郁，商业总面积约60 ha.，占整个新城总建筑面积的三分之一。

规划路网与周边路网相谐调，原规划中穿越新城的"十"字形城市主干道在新规划中予以保留，新规划中支路出口均与周边路网接通，同时地块划分尽量与周边地区的地块划分大小一致，以保证城市脉络的延续性。

用地面积：	98 ha.
总建筑面积：	2 200 000 m²
建筑密度：	30%
容积率：	2.2

The planning of Harbin Aijian New City community was an influential international competition. KFS placed first for the urban planning. Eight years of construction followed with 90% of the floor area completed. This included a series of landmarks and a large amount of residontial floor space.

Aijian New City community is situated at the bank of Songhua River. The site's former owner was the Harbin Locomotive Factory which is the largest and the first of its kind in China.

New concepts of urbanism are introduced using elements that elevate cultural and aesthetic levels. The design respected the local urban texture and established continuity within the urban spaces and their Harbinese personality. A modern architectural language combined with local character was applied to the theme of the site. Environmental protection acted as a core principle and ensures the making of a green and vibrant new community.

Open spaces were created among dense residential areas by transforming old plants into ellipse-shaped green parks. Condominiums were planned around these parks, and defined the centre of the community. The high-rise buildings, outside of the commercial ring, faced the park, offered the pleasant view to theirs owners.

The commercial part planned with a system that limited the building type to either low-rise or multi-storey buildings were evenly spread along the street created an inviting atmosphere. One-third of the total site area, or 60 ha., was used for commercial activities.

Precautions were taken during the planning process for the roads to keep the correlation with the city road network. The branches of the main traffic corridor, have been kept from a previous planning stage, were linked to the road network. The new building blocks retained the similarity of the urban texture, and the dimensions and scale of the existing buildings.

Site area:	98 ha.
Gross floor area:	2,200,000 m²
Building density:	30%
F.A.R:	2.2

海南三亚亚龙湾一号规划设计
1# Yalong Bay Planing, Sanya, Hainan

中国，海南，三亚 Sanya, Hainan, China
业主：海南申亚置业有限公司
Client: Hainan Shenya Group
设计时间 Design: 2013

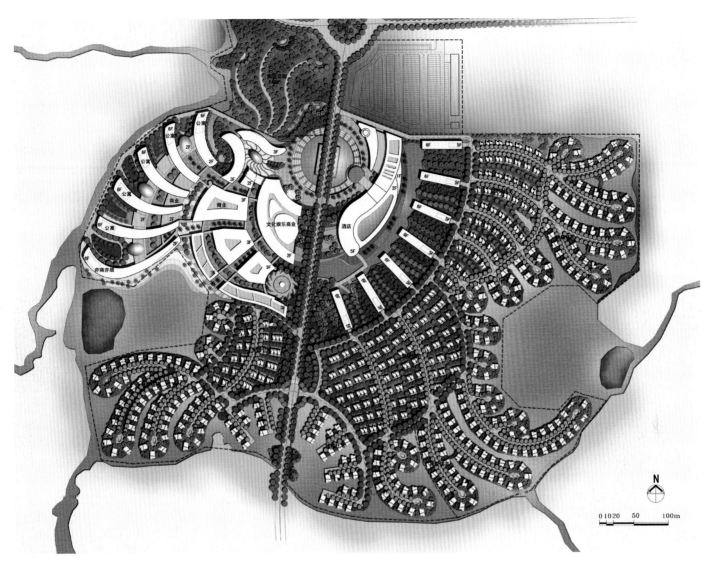

总平面图　Site Plan

海南三亚亚龙湾一号项目位于三亚亚龙湾，与周边的多个五星级酒店相邻，地块位于进入亚龙湾的主干道椰风路(龙溪路)的两侧。

基地的一侧朝向大海，两边背靠青山，其中包含着著名的电影《非诚勿扰》的拍摄基地和电影中出现的悬索桥，在椰风路上也形成极具视觉冲击力的城市门户。其中北侧的广场将酒店和商业两大主体紧密连接成为一个整体，化解了椰风路两侧建筑相隔较远的不利因素。

整体的设计思路依循中国传统的"凤凰攀枝"的理念而形成。设计将五星级酒店、商业建筑和居住建筑融为一体，在规划中表现了传统的吉祥概念。不论是平视还是从附近山体鸟瞰，其独特的建筑形式必将成为当地的城市标志。

三大片区分别为五星级酒店、大型商业综合体及高档度假社区。

The site is located on Yalong Bay in Sanya, Hainan. beside about 20 five star hotels.

The site is facing the sea with two sides adjacent to the mountain. There is a view of the famous suspension bridge on the site where the popular movie "If You Are the One" was filmed. The plaza on the north combines the two commercial zones to form a vivid portal city and solves the issue of the building's separation on both sides of Yefeng road.

The overall master plan follows the traditional Chinese concept of a 'phoenix climbing branches.' This symbolizes the fortune of combining five-star hotels, commercial and residential buildings. The unique design which can be either viewed horizontally, or from birds view will become an iconic feature for the city.

Three areas are composed of a five-star hotel, a large commercial complex, and an upscale resort community.

总用地面积：	6.76 ha.
总建筑面积：	57 000 m²
容积率：	0.84

Site area:	6.76 ha.
Gross floor area:	57,000 m²
F.A.R:	0.84

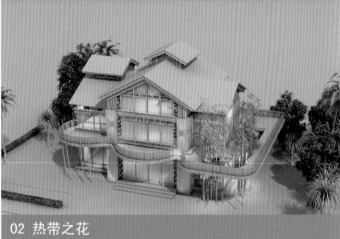

01 丛林别墅

02 热带之花

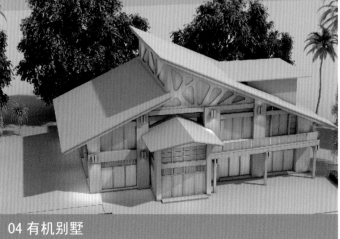

03 新南洋别墅

04 有机别墅

05 海岛风情

06 南洋风情

07 金枝玉墅

08 盈灿别墅

09 岭南别院
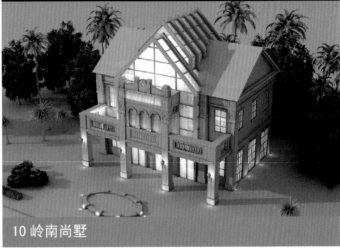
10 岭南尚墅

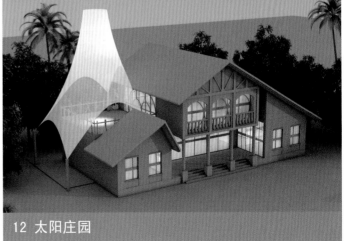
11 岭南峻墅
12 太阳庄园

13 斯特拉斯别墅

14 加州别墅

15 摩登 LOFT

16 钻石别墅

17 渔石别墅

18 美式风情

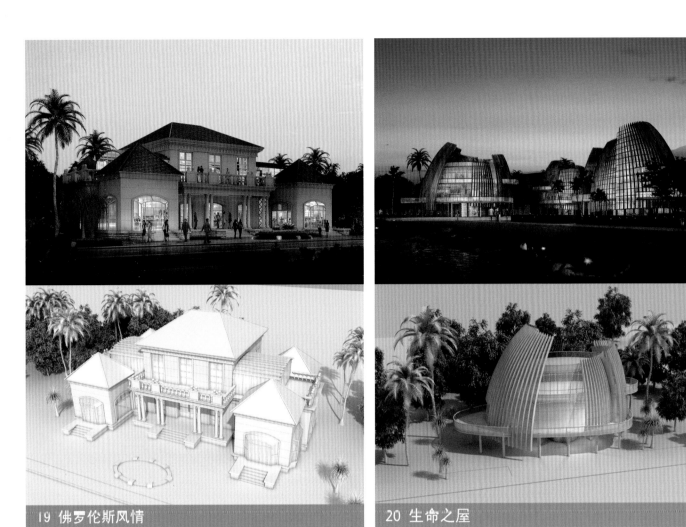

19 佛罗伦斯风情   20 生命之屋

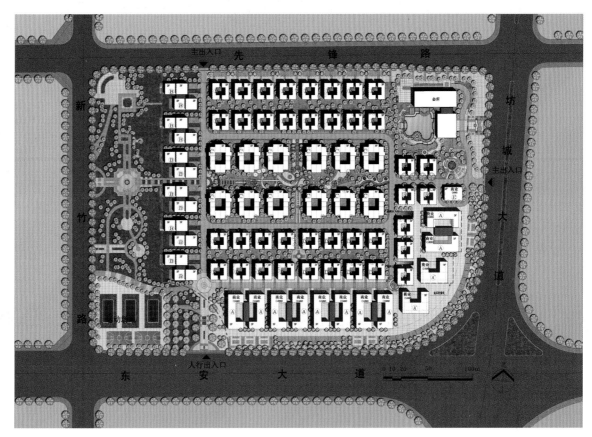

总平面图　　Site Plan

长泰锡东国际社区位于江苏省无锡市锡山经济开发区，北临先锋路，南靠东安大道，西侧为新竹路，东侧为坊城大道。地块用地性质为商业办公混合用地。随着锡山区被规划为新的城市中心区，加之高铁站与本案基地距离较近，因此该区域势必将成为无锡市的重要形象窗口。商业与办公混合的形式也必将形成新的城市综合体和繁荣点，建成后将为该区域以及整个锡山区带来24小时的活力。

方案构思来源于基地周边自然与现代相结合的现状。作为展示无锡市的重要窗口，项目必将充分地展示出强烈的现代气息以及严谨并兼具无限活力的状态。设计师通过"街道"概念以及亲人尺度的引入，将整个基地内建筑化整为零，消解了商业办公中心区可能带来的与生活尺度的脱节以及与传统城市肌理的割裂。在基地外围沿东安大道以及新竹路一侧城市公共绿化的两条边界人了兼用对外和对内双重功能的商业景观步行带，两侧建筑通过小体量的灵活布置，与无锡市传统城市空间中的尺度相近，成功营造出宜人之感，既融合了办公需要的宁静与亲近之感，同时又极大地增强了作为商业所展示出的购物时的舒适性，使整个界面呈现出一种舒适安详而又充满活力之感。

立面设计整体采用简化的ART-DECO（装饰艺术）风格，纵向线条与极具现代感的简洁风格相结合，使建筑既具有古典的幽雅，又流露出强烈的现代感。立面采用红砖与仿石材喷涂相结合，将近观的精美细部与远观的整体大气完美结合，在各个沿街面上都呈现出极高的建筑品质。

Changtai International Community is located at Wuxi city of Jiangsu province, surrounded by Xianfeng road, Dongan avenue, Xinzhu road, and Fangcheng avenue. It is an office and commercial area. Xishan area in which the project is located is a newly developed area which plays a role as a city center; thus the site is of great importance to the city. The office and retail is designed in a way that it provides liveliness and energy to the residents throughout the whole day.

The concept of this project is based on the combination of modernity and nature of the area. Seeing the importance of the location, the architects used the idea of "streets" to minimize the volumes of buildings thus creating more pedestrian-friendly ambience and also harmony with the existing urban texture. Appropriate landscape design and flexible building volume arrangement is applied to create more connection between architecture and its users, providing the privacy that an office building would need and the openness of retail buildings, fulfilling both needs while creating an energetic and flexible space.

Facade design is of simplified art-deco style which emphasizes on vertical lines to create a mix of modern and classical look. Material used in the facade is red brick and faux stone paint in an elegant combination, making the buildings stand out in the important location of the city.

总用地面积约：	100 603 m²
地上建筑面积：	77 254 m²
容积率：	0.7

site area:	100,603 m²
Total Building area:	77,254 m²
F.A.R:	0.7

西昌邛海凤凰谷度假小镇
Phoenix Resort Valley, Qionghai, Xichang

中国，四川，西昌 Xichang, Sichuan, China
业主：成都大天实业发展有限公司
Client: Chengdu Datian Industry Co.,Ltd.
设计时间 Design: 2010

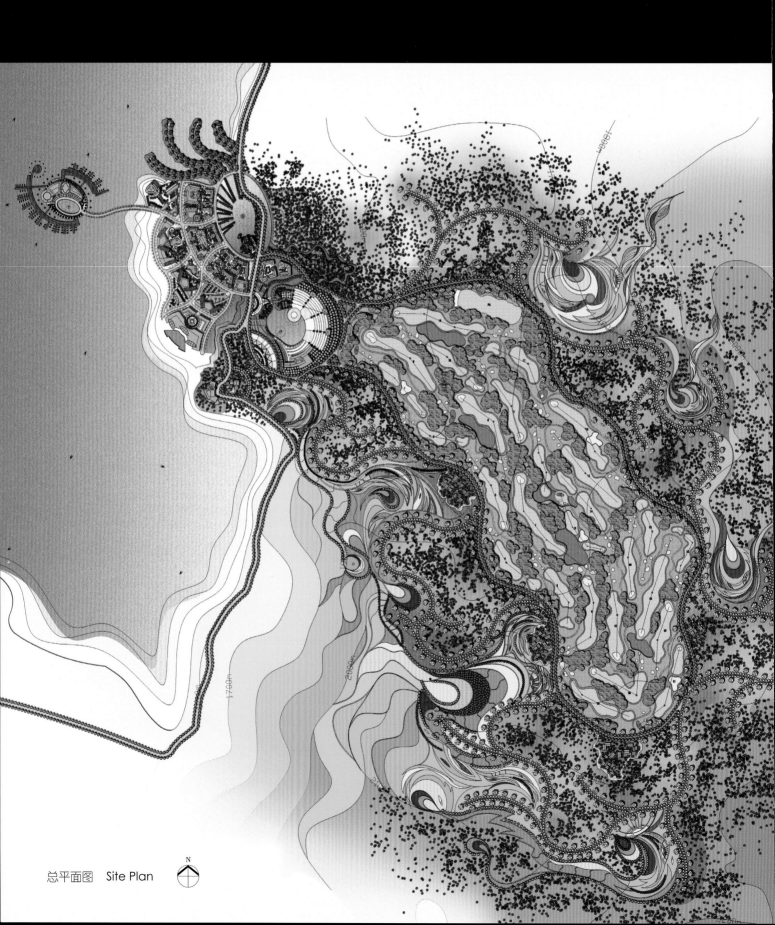

总平面图　Site Plan

西昌邛海青龙寺凤凰谷度假小镇位于西昌邛海泸山风景区，邛海的东岸青龙寺片区。基地依山傍水、风景秀丽，是集五星级酒店、国际商务会议、养生、休闲、居住、民俗体验与时尚运动于一体的复合型国际湖滨山地高尔夫休闲度假区。

项目基地山势起伏绵延，整个用地呈一个扁长的"盆地形"，边高中低，仅在西侧滨湖区打开一个缺口。设计上结合地形地貌，将度假区规划为三大板块：湖滨度假酒店区，风情小镇与山地高尔夫度假区。

项目以崇尚自然，回归生态为设计理念，结合基地地形构筑成"凤凰谷"之大自然绿色生态景区的主旋律。

无论是人工景观还是水体沙滩都是结合现状、融入文化、顺应自然，体现生态的景观价值观。

1. 邛海——开阔的水景面，加大商业及公共广场景观的布置。

2. "彩虹梯田"——作为区域内最主要的景观之一，为项目整体规划之"凤凰谷"勾勒出动人的一笔，使人们最大限度地与自然接近，与绿色相拥，创造并保留一片宁静的生态彩色农作物梯田。

3. "绿色原生态"——穿梭于山水之间的自然材料的色彩和质感肌理，创造出人性化、有益身心的色彩环境，塑造休闲、健康、和谐和多元化的生活度假区。

基地面积：	277 ha.
建筑面积：	110 000 m²
容积率：	0.4

The Phoenix Resort is situated in the Qionghai lake tourist zone of Xichang with the Green Dragon Temple on the mountain side on the east shoreline of Qionghai Lake. The joy of the great view enhanced by the support functions, of a five-star hotel, business conference centre, spa, leisure, living, local cultural experience, golf and other trendy sports offer a superior life style.

The surrounding mountains mirror the elongated shape of the site, making a basin-like gesture with an opening on the west bank, followed by the separation of the site into three parts, the lake-side resort hotel, a romantic town and hilly golf course.

The project advocated the importance of nature by imposing the back-to-nature concept as the most prominent design guideline. Environmental elements of water, landscape and vegetation clusters formed variable seasonal views.

The man-made landscape represented a solution of the natural conditions for the design concept. This design consisted of three main concepts.

1. Qionghai—The elements of wide open waterscape, large commercial volumes and public spaces.

2. "The escalated rainbow fields"—The cooperation of the fields with the peripheral landscape enlarged the Phoenix pattern. The vegetation of a distinctive color pallet mimed the feathers of the Phoenix, bringing people together in nature in harmony.

3. "The green original ecology"—Using the colors and textures of the mountain and the water as materials and a personifying a healthy therapy environment of colors was created.

Site area:	277 ha.
Gross floor area:	110,000 m²
F.A.R:	0.4

Yu Mountain Shang Lake Eco City, Changshu

中国，常熟 Changshu, China
业主：上海长甲置业有限公司
Client: Shanghai Changjia Real Estate Co.,Ltd.
设计时间 Design: 2010

虞山—尚湖的历史，是一个圈湖造地的历史。规划设计在不断反省，尊重自然，以自然为师，注重自然生态的基础上还地于湖。充分利用尚湖地区自身水资源，创造独具特色的滨水社区。从虞山山顶望下，本项目如尚湖蔓延出的朵朵"水上花"。

常熟虞山—尚湖水中花规划用地约300 ha.，位于尚湖南岸。用地南接南三环路，东接元和路，本身属国家4A级风景区，周边有多处成熟的景区和相关配套工程。规划结构上，加强尚湖生态廊道和环湖路的联系，形成完整便捷的交通框架，让人们在第一时间欣赏到尚湖和虞山的主体生态美景，并自然而然地串联起各大片区。

生态之城：尚湖是常熟生活饮用水的重要水源地之一，规划保留原有"湿地鸟岛"等宝贵的生态资源，坚决杜绝一切对水体可能有污染的水上活动。整个规划像从尚湖自然生长出来的"水上花"。

休闲之城：结合现有度假村及水街商业，规划更加完善的滨湖度假配套设施，并以沙滩、木栈道等亲水形式丰富尚湖驳岸，使人们可以与尚湖零距离接触。

文化之城：从虞山—尚湖景区到沙家浜景区，常熟从来不缺历史文化典故。规划在滨湖沿岸设想增加国际会议中心、水上音乐厅等市政综合配套设施以及水幕电影、许愿台等滨水设施，让游客享受从听觉、视觉到心灵的全方位的文化体验。

养生之城：养生住宅、康体中心、水疗会所、高尔夫等设施，使惬意的生活无处不在。

水上之城：水网发达是江南水乡的普遍优势。规划保留用地内大部分宽阔水域，并进行梳理，形成一座魅力四射的水上之城。

总用地面积约：	300 ha.
建筑面积约：	1 300 000 m²
容积率：	0.4

Historically Yu Mountain – Shang Lake, land reclamation caused progressive damage to the environment. The finding of a more ecological acceptable and friendly way of development is today of great importance. The design is predicated on the site's water resources in the planning process to create a splendid lake-front community.

Ecological City: Shang Lake is not only the main water source for the local communities, it is in addition the preservation area of the original "Wetland Bird Island". All actions that could potentially pollute the lake are strictly forbidden. The design implications that follow from this restriction have led to a scenery of buildings shaded by trees in fields of flowers.

Leisure City: The concept is the combination of resorts with commercial functions along the lake that make for a perfect lakeside holiday experience. This experience is enhanced by the design of the public space and public furniture, of beach and the wooden docks.

Cultural City: The concept is the whole Changshu Yu Mountain area, stretching from Shang Lake Resort to Shajiabang Resort. It has a historical significance and KFS have included an international conference centre, a concert hall and other public functions in the master plan. The public space will be decorated with curtains of water, wish fountains and other typical waterfront features. This modern touch to the atmosphere of the lakeside enables the tourists to enjoy the full scope of cultural experiences.

Healthy City: The concept at any given location on the side it will be comfortable to live, due to the access to supporting functions, such as apartments, a recreational and sports center, a SPA and golf clubs.

Water City: The concept is the water network, well developed and used as a general advantage for Jiangnan. In the master plan the water retain a wide and open character to form the backdrop for the glamorous water-city.

site area:	300 ha.
Total Building area:	1,300,000 m²
F.A.R:	0.4

COMMERCIAL DESIGN

公共建筑

上海海上文博苑方案设计
Haishang Cultural Museum Complex, Shanghai

中国，上海，嘉定 Jiading, Shanghai, China
业主：上海市嘉定区规划和土地管理局
Client: Shanghai Jiading Urban Planning Administration Bureau
设计时间 Design: 2015

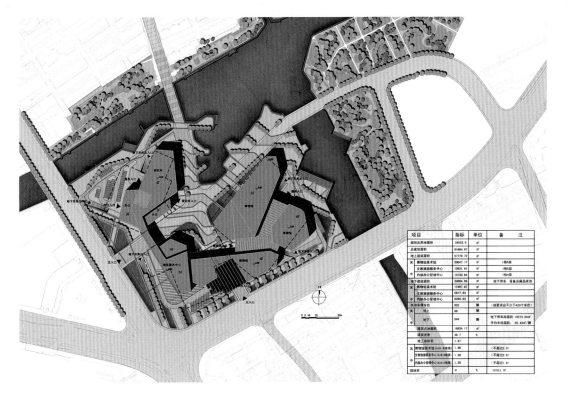

总平面图 Site Plan

上海文博苑位于上海嘉定老城区南端,是一个集文化、旅游、休闲和园林为一体的创新文化旅游博览园,主题功能包括博物馆和美术馆。

方案构思来源于巨石从天而降,撞击地面释放出无限能量及光芒,将之喻为人类之精华和瑰宝并演绎为"嘉定之门",创造建筑与水体之间的和谐灵动,打造独特的"江南艺术水乡"。博物馆整体造型简洁有力,富于雕塑感,仿佛一颗颗宝石镶嵌在江南韵味的山水之间。

三座建筑均由城市道路经由提供大量人流聚集的入口广场进入主入口大厅,再经垂直交通到达建筑内各处;次入口为博物馆工作人员和货运后勤使用。访客由西侧的入口大厅步入中心的共享中庭,可围绕中庭参观面积不等的多个展览空间。藏品由地下一层到达库区,通过专用货梯进入各展厅。

总用地面积约: 3.5 ha.
地上建筑面积: 81 864.7 m²
容积率: 1.67

The site is located in the northern part of Jiading old town, Shanghai. The program contains culture, tourism, leisure and garden four aspects with museum and art gallery as the main function.

The design concept is inspired by the falling stone from sky which reveals its treasure when hits the ground and becomes the gate of Jiading. The geographic location of this city which has the unique "Jiangnan water village" contributes with the elements to create a water landscape in the site that co-exists with the buildings generating an iconic landmark for the whole city.

The visitors could enter all three buildings from the main entrance in the central core where visitors gather together then travel vertically to different buildings. The secondary entrance is for museum staff use of internal flow and logistic transportation. Visitors enter the main hall from the west and explore the different exhibition spaces which is arranged around the atrium. The collection is transferred by private ladder from the basement to the display area.

site area: 3.5 ha.
Total Building area: 81,864.7 m²
F.A.R: 1.67

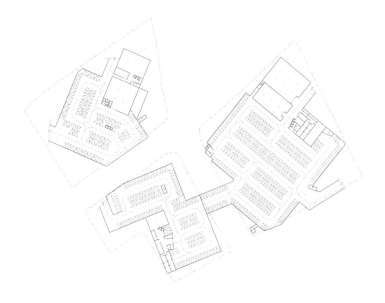

地下一层平面图
B1 Floor Plan

一层平面图
1st Floor Plan

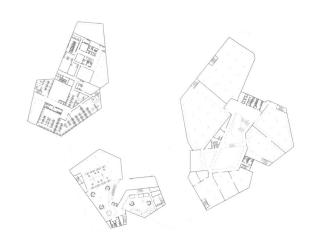

二层平面图
2nd Floor Plan

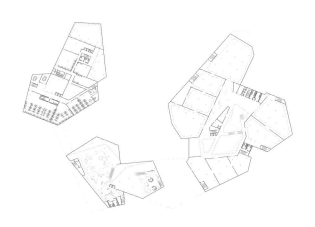

三层平面图
3rd Floor Plan

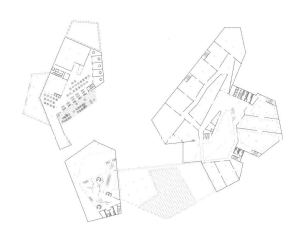

四层平面图
4th Floor Plan

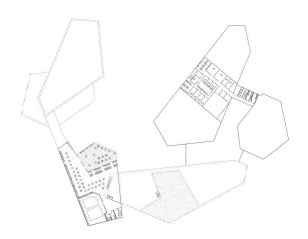

五层平面图
5th Floor Plan

国际乒乓球联合会博物馆设计方案
Proposal for International Table Tennis Museum, Shanghai

中国，上海 Shanghai, China
业主：上海体育学院
Client: Shanghai University of Sport
设计时间 Design: 2014

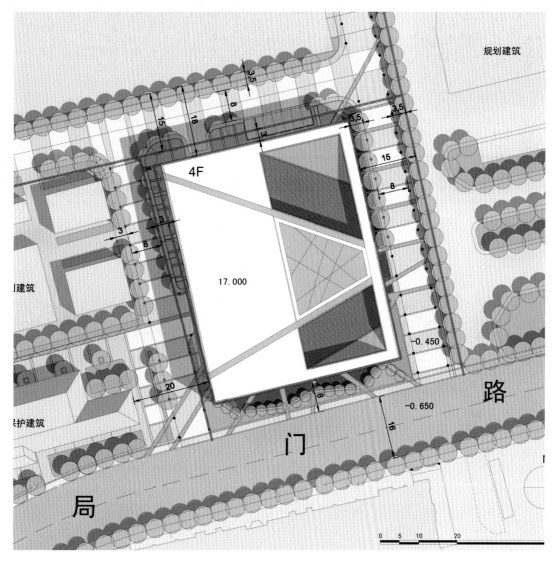

总平面图　Site Plan

国际乒乓球联合会博物馆位于生气勃勃的原2010上海世博会园区，是黄浦江畔关键的艺术及历史保护区。此处拥有不少的重要博物馆。其中15-02地块为将建成的世博会博物馆。

方案构思来自乒乓球从小游戏演变到国际化竞赛项目的演变过程，同时也考虑到它是提供大众观看的一项运动。借鉴这样的想法，建筑里包含了不少的给观众观看和参加游戏的公共空间。建筑的材料选择有所讲究，比如，中庭里的斜坡选用的是类似乒乓球拍的胶合板，观众通过透明玻璃进行观看，草皮给观众们提供亲切的公共交流空间。

博物馆整体造型简洁有力，富于雕塑感，折叠起伏的墙面代表了乒乓球运动的节奏与速度。玻璃幕墙与金属穿孔板的结合设计赋予了建筑现代感的同时也含蓄而内敛地表达了对乒乓球运动的热爱。入口处折叠翻转的外立面在提供了内外空间互动交流的同时也形成了一个标志性的外展空间和入口意向。

The site of the new Table Tennis Museum is located in the site of 2010 world expo, a few meters from the river. The West Bund area is an important artistic and historical preservation center. Also, many famous museums are established in this area. The neighboring property (15-02) is the World Expo Museum.

The concept of the design is based on the transformation of pingpong from local to international, and the growing size of the participants, which is the players and the spectators. With this, the project has included interactive spaces that would bring the visitors into the game. The choice of material is carefully considered to symbolize the game itself, with the wood decking being inspired from the laminated wood of the paddles. Large surface of glass enables the visitors to witness the activities in the courtyard as an spectator would witness a pingpong match.

Building form symbolizes the movement and bouncing of a pingpong ball. Curtain wall and semi transparent metal sheet imbues the look of a modern style building and reflects the spirit of the game. The dynamic volumes of the exterior has created a connection between interior and exterior space, enabling the visitors on the second and third floor to have a view to the entrance courtyard.

总用地面积约：	5 000 m²
地上建筑面积：	10 338 m²
容积率：	1.68

site area:	5,000 m²
Total Building area:	10,338 m²
F.A.R:	1.68

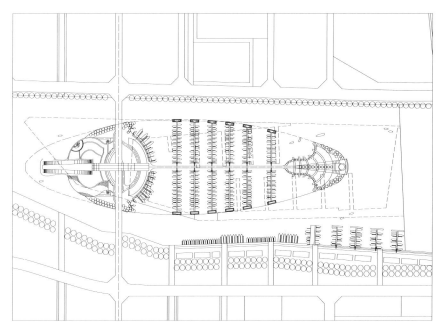

总平面图　Site Plan

珠海横琴岛横琴大酒店位于珠海市横琴岛，东临澳门大学，与澳门一河之隔，南北临大小横琴山脉。其靠山临海，景观环境优越。

设计立意为横琴—情岛。横琴毗邻港澳，这里原来是两个岛：大横琴岛和小横琴岛。因其地形和山势，大小二岛像横在南海碧波上的的两架古琴，千万年来，日日夜夜和着山风与海涛弹奏着山之歌、风之歌与海之歌。

设计为迎合其地理位置特色及周边环境特征，取其横琴寓意作为建筑基本形式。

总用地面积：	3.5 ha.
地上建筑面积：	68 000 m²
容积率：	1.90
建筑高度：	82 m

The site for the hotel is located at Hengqin Island in Zhuhai with Macau University on the east, and is separated from Macau by a river. On the north and south of the site there is Big Hengqin and Small Hengqin mountain range. This geographical feature provides the site with stunning landscapes.

The design idea is inspired by Hengqin, literally meaning "horizontal lute"—Island of romance; being adjacent to Hong Kong and Macau, it was originally two islands: Big Hengqin and Small Hengqin. These two parallel islands resemble a couple of Chinese ancient lute, eternally playing music for the mountain, the wind, and the sea throughout the day.

Hengqin Hotel combines all these geographical features to result in a form of a horizontally positioned Chinese ancient lute.

Site area:	3.5 ha.
Above ground G.F.A:	68,000 m²
F.A.R:	1.90
Building height:	82 m

上海浦东锦绣华城假日酒店
Holiday Inn Pudong, Shanghai

中国，上海 Shanghai, China
业主：上海大华集团
Client: Shanghai Dahua Group
设计时间 Design: 2005
建成时间 Completion: 2010

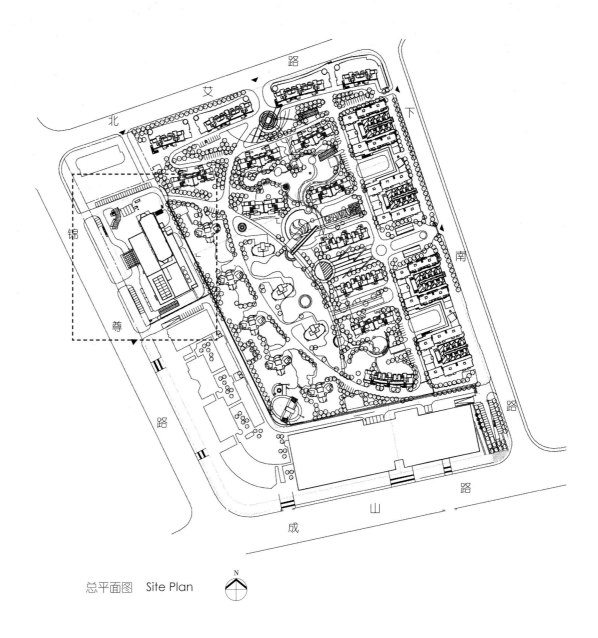

总平面图 Site Plan

上海浦东锦绣华城假日酒店位于上海浦东新区，属于锦绣华城社区的11#地块。该地块西面紧临锦尊路，南面与北面皆为商业建筑设施用地，东面为住宅区。地块东侧靠近轨道交通站点，锦尊路东侧即为湿地主题公园，整个酒店也以体育公园为主要朝向，使最多的房间能面向最佳的景观。而南面的商业设施与酒店互相烘托人气，方便使用。

本案地块呈条状，酒店区域用地面积为1.1 ha.，酒店总建筑面积为41 000 m²，总高21层；酒店地下层主要为酒店车库和管理用房；4~21层为酒店客房区，客房数约为219套；1~3层主要设有24小时餐厅、特色餐厅、500人宴会厅、高端会议室、游泳池、水疗等休闲娱乐设施。便捷的交通与商务人士高效率的生活节奏十分合拍，各种高端设施满足了客户的休息娱乐要求，使客户充分享受酒店的舒适性。

The Holiday Inn site is located in the Pudong New District of Shanghai, and belongs to The Graceful Oasis City. The hotel is west of Jinzun Road, south and north of the commercial facilities and east of the residential area. A theme park is positioned in the east. The best view, for the hotel is facing the theme park. The commercial functions and the hotel ensure an active character and popularity for the area.

The 1.1 ha for the 21-storey hotel site is an elongated 'L' shape. The gross floor area is 41,000 m². The parking area and offices of the hotel are located in the underground. The hotel occupies the 4th floor to 21st floor with 219 hotel units of standard rooms and suites. The hotel provides a 24-hour restaurant, a theme restaurant, a 500-seats banquet, conference rooms, swimming pool, SPA and other recreational facilities. The convenient circulation suits the fast pace of a businessman's lifestyle. The comfortable designed facilities will satisfy the clients' needs.

用地面积：	1.1 ha.
总建筑面积：	41 000 m²
容积率：	2.7
酒店地上建筑面积：	32 000 m²
酒店客房数：	219

Site area:	1.1 ha.
Gross floor area:	41,000 m²
F.A.R:	2.7
Above ground G.F.A of hotel:	32,000 m²
Hotel rooms:	219

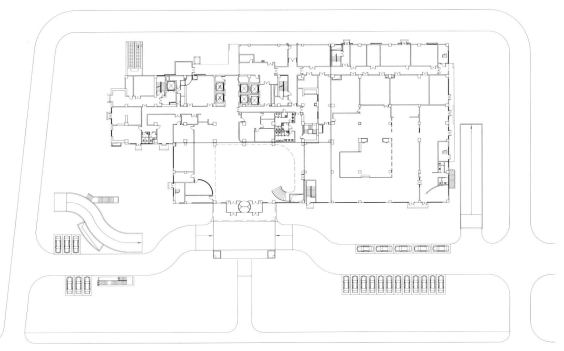

底层平面图　1st Floor Plan

二层平面图　2nd Floor Plan

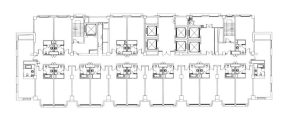

标准层平面图　Typical Floor Plan

无锡千禧大酒店
Millennium Hotel, Wuxi

中国，无锡 Wuxi, China
业主：无锡鑫畅置业有限公司
Client: Wuxi Xinchang Real Estate Development Co.,Ltd.
设计时间 Design: 2006
建成时间 Completion: 2009

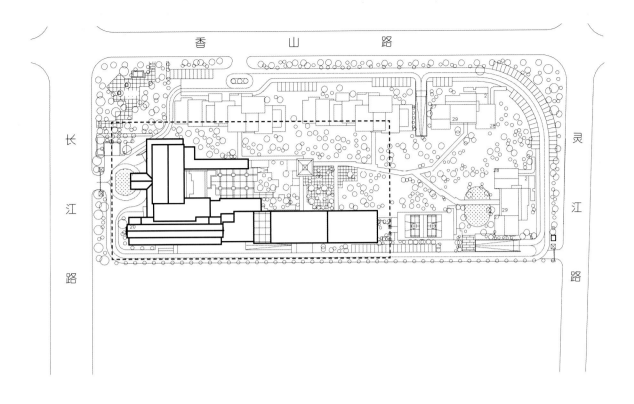

总平面图　Site Plan

无锡千禧大酒店位于无锡市无锡新区内，西侧是长江路，北侧与香山路连接，作为整个基地的一部分。共计两栋建筑，地上建筑面积约为35 000 m²。

酒店区主要布置在基地南部。规划在西侧长江路设置酒店主出入口，主出入口南侧靠南界线处设酒店后勤机动车入口，两个出入口与相邻公寓区的主入口互不干扰。地下室主要为机动车库及设备用房，还有少量娱乐休闲及员工服务用房；一、二层公共建筑主要包括健康运动功能、餐饮娱乐功能、办公会议功能等；三层~二十二层为酒店客房，共设置客房306间。

酒店建筑拥有一个约40 m宽、150 m长的景观花园，可通过半封闭式的"景观回廊"到达花园以及基地内其他区域的各栋建筑，使酒店能更好地为不同人群进行服务，提高了利用率。同时，通过"景观回廊"形成各主题景观空间，体现了中式园林"移步异景"的特色。

在设计中，KFS坚持建筑、室内的一体化设计，完成了酒店所有的建筑设计和室内设计，使室外、室内空间相得益彰、互相交融。

用地面积：	1.1 ha.
总建筑面积：	35 000 m²
容积率：	2.9
酒店客房数：	306

The design is, as part of a larger developed site, located in the Wuxi New Area in the city of Wuxi. The hotel is east of Changjiang Road and south of Xiangshan Road. The site area is 1.1 ha and supports two hotel buildings with a combined gross floor area of 35,000 m².

The main hotel entrance is located at Changjiang Road with the vehicle entrance to the south. The entrances of the apartment blocks are set apart from the hotel entrances. The lower level of the hotel provides secure underground parking places, equipment rooms, recreational rooms and service rooms. The gyms, restaurants and conference rooms are located on the first and second floor. The hotel offers 306 standard rooms and suites.

The buildings surround the inner garden square. This architectural design feature provides privacy to the 40 m wide and 150 m long inner garden. The inner garden enables an easy circulation amongst the different buildings. The landscape garden in the middle is surrounded by a landscaped corridor, which makes for a satisfying stroll through natural gardens

KFS provided and integrated full design service for the interior and exterior design.

Site area:	1.1 ha.
Gross floor area:	35,000 m²
F.A.R:	2.9
Hotel rooms:	306

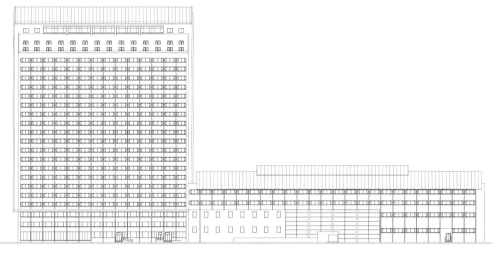

南立面图　South Elevation

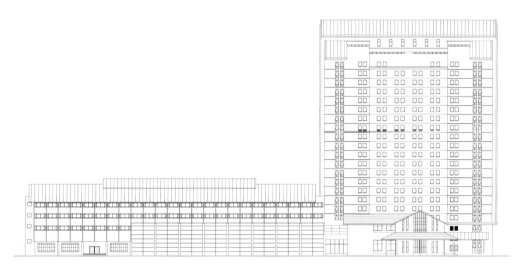

北立面图　North Elevation

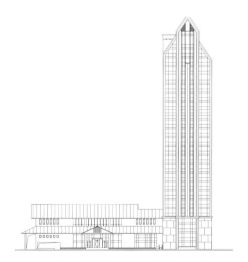

西立面图　West Elevation

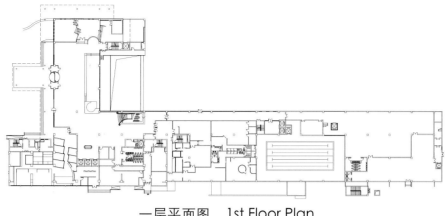

一层平面图　1st Floor Plan

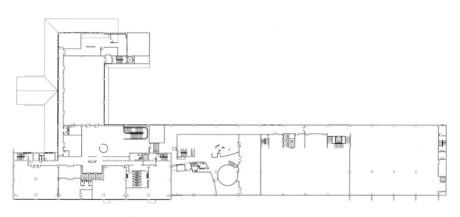

二层平面图　2nd Floor Plan

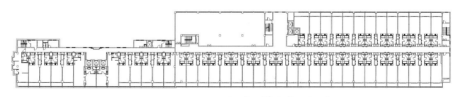

三~四层平面图　3rd~4th Floor Plan

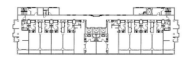

标准层平面图　Typical Floor Plan

成都新东方千禧大酒店
Millennium Hotel, Chengdu

中国，成都 Chengdu, China
业主：成都新东方置业有限公司
Client: Shangfang Group/New Oriental Real Estate Inc.
设计时间 Design: 2004
建成时间 Completion: 2009

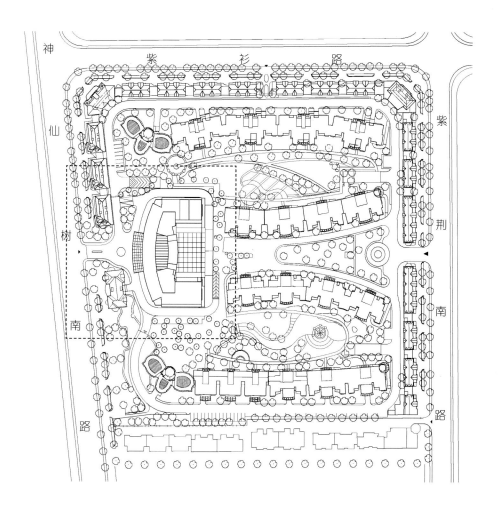

总平面图　Site Plan

成都新东方千禧大酒店位于成都市南部，神仙树路东侧的高新技术区内，为一家高标准的五星级酒店。

酒店主楼共14层，裙房2层，内设客房319间，同时有商务、会议、餐饮、休闲等多种功能设施，是成都的高标准、多功能的标志性五星级酒店。

基地位于"上海花园"内，并紧邻城市道路。酒店西南处有2 000 m² 左右的市政绿地，为酒店提供了良好的景观。酒店的设计考虑到最大地利用景观资源，有80%的房间可以看到绿地。内部设计有一个底层至11层的大型中庭共享空间，它配备有独立的车行和人行系统以避免周边居住区的干扰。新古典主义的建筑风格采用了简洁的处理以使之与周边的现代建筑相谐调。

用地面积：	0.8 ha.
总建筑面积：	38 000 m²
容积率：	3.7
酒店客房数：	319

The design is for a five-star hotel in a high-technology zone east of Shenxianshu Road, South in Chengdu, in Sichuan Province.

The tower building is fourteen storeys, with two podiums and one underground level, with 319 suites, several conference rooms, restaurants and entertainment facilities. The tower is designed to be a high quality, multi-use hotel in Chengdu.

The tower is situated inside "Shanghai Garden" and enclosed by streets. The 2,000 m² City Park is south of the tower. The hotel has some great views of the surrounding landscape. This quality has been the guiding principle for the orientation of the suites, of which eighty percent face the open green area. The different traffic systems as well as the functional systems are designed to be separate from the nearby residential activities. The neo-classical style of the hotel, with a 11-storey high atrium, forms the harmonious link between the modern style apartment buildings.

Site area:	0.8 ha.
Gross floor area:	38,000 m²
F. A. R:	3.7
Hotel rooms:	319

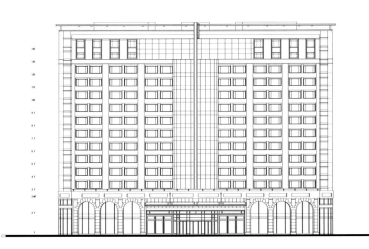
立面图　Elevation

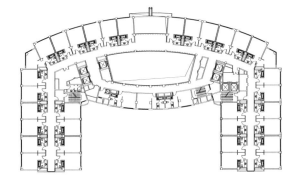
标准层平面图　Typical Floor Plan

珠海横琴岛中大金融大厦
Zhongda Financial Tower, Hengqin, Zhuhai

中国，珠海 Zhuhai, China
业主：中大控股
Client: Zhongda Group
设计时间 Design: 2014

项目位于珠海横琴新区十字门中央商务区十字门大道东侧、汇通五路南侧、荣港道西侧、汇通三路北侧。用地靠近离岸金融岛中心绿地，东北远眺澳门及氹仔。

项目地理位置与澳门隔海相望，基地海景优越。本项目有一栋250 m高的商务办公楼和两栋100 m高的商务公寓楼及商业裙房所组成。

总用地面积：	1.87 ha.
总建筑面积：	111 000 m²
容积率：	5.80

The project is located at Shizimen central business district of Hengqin Area of Zhuhai. It is surrounded by Shizimen avenue on the east, Huitongwu road on the south, Ronggang avenue on the west, and Huitongsan road on the north. It is close to Li'an Financial Center, and at the far northeast is Macau and Hanzai.

The geographic location with Macau across the sea provides an outstanding ocean view. The project combines a 250 m commercial office tower, two 100 m commercial apartment and its podium buildings.

Site area:	1.87 ha.
Gross floor area:	111,000 m²
F.A.R :	5.80

裙楼一层平面图
podium 1st Floor Plan

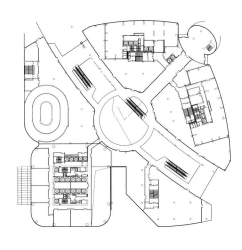

裙楼三层平面图
podium 3rd Floor Plan

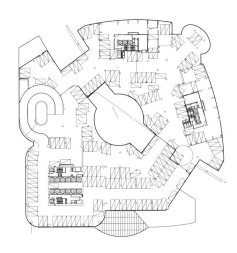

裙楼四层平面图
podium 4th Floor Plan

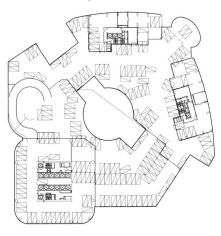

裙楼五层平面图
podium 5th Floor Plan

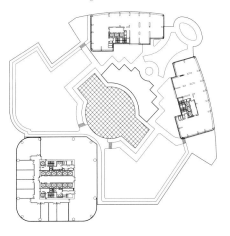

裙楼六层平面图
podium 6th Floor Plan

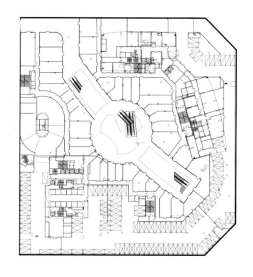

地下一层平面图
Basement 1st Floor Plan

地下二层平面图
Basement 2nd Floor Plan

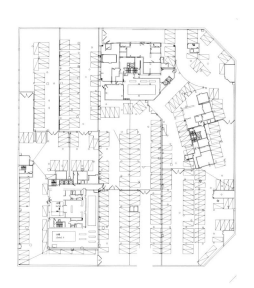

地下三层平面图
Basement 3rd Floor Plan

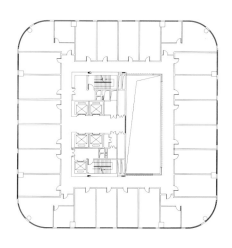

商业办公楼典型平面图
office typical Floor Plan

中国，上海 Shanghai, China
业主：上海金缔联创置业有限公司(长甲集团)
Client: Shanghai Jindi Real Estates Co.,Ltd.
设计时间 Design: 2010
建成时间 Completion: 2014

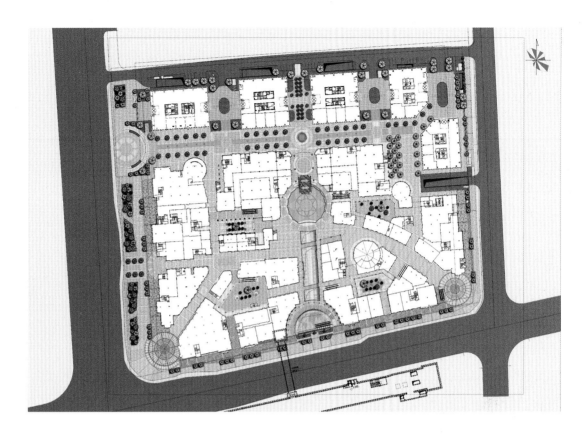

总平面图　Site Plan

本工程位于上海浦东张江地区，南临祖冲之路，西靠金科路，东侧为哈雷路，北至相邻地块地界。总用地面积8.2 ha.，北部为高层办公楼，南部为商业购物广场。

商业广场建筑层数为2~4层，外观设计以整齐简洁的新古典风格为主，局部配以现代风格立面体量及元素，体现传统与现代的碰撞与融合。在这里，古典的典雅高贵与现代的激情时尚同时呈现，符合现代商业多彩多姿的发展方向。

总用地面积：　　　　　　　8.2 ha.
总建筑面积：　　　　　　　165 000 m²

This project is located in Pudong Zhangjiang area of Shanghai, with Zuchongzhi Road, Jinke Road, and Halei Road as the surrounding streets. The project encompasses 8.2 ha in total, with high rise office buildings on the north and a commercial square on the south.

Design of the commercial square utilizes both neoclassical and modern style in the facades, displaying a fusion of modern and tradition. The neoclassical facade of the commercial space include buildings with a variety of structural and facade elements, providing the customers with an array of experiences through changes in ambience during their leisure activities. The elegance of the classical combined with the lightness of modern style results in a tasteful yet flexible characteristic that define this highly used commercial space.

Site area:　　　　　　　　8.2 ha.
Gross floor area:　　　　　165,000 m²

海南三亚亚龙湾一号商业综合体
1# Yalong Bay Commercial Building, Sanya, Hainan

中国，海南，三亚 Sanya, Hainan, China
业主：海南申亚置业有限公司
Client: Hainan Shenya Group
设计时间 Design: 2013

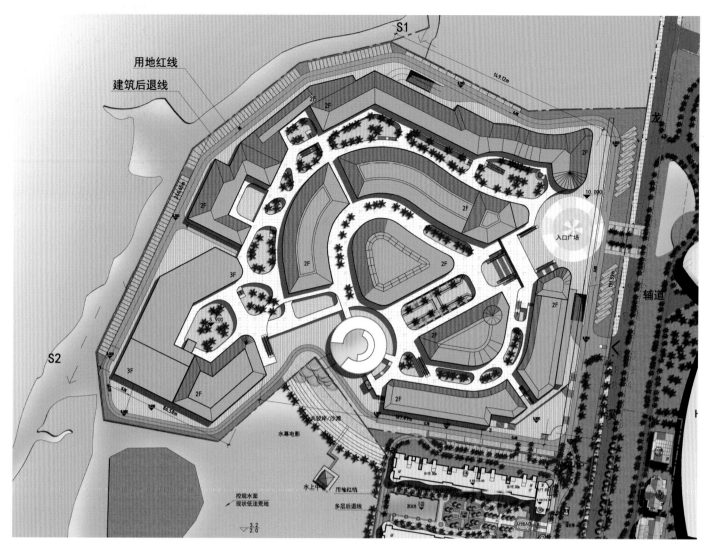

总平面图　Site Plan

海南三亚亚龙湾地块（亚龙湾一号项目）位于三亚亚龙湾，与周边的多个五星级酒店相邻，地块位于进入亚龙湾的主干道椰风路(龙溪路)的两侧。

两大门户分别位于项目南北两端，是椰风路上极具视觉冲击力的两大城市门户，其中北侧的广场将酒店和商业两大主体紧密连接成为一个整体，化解了椰风路两侧建筑相隔较远的不利因素。

整体商业建筑在考虑安全舒适及方便购物的基础上适当地引入了当地的传统戏剧文化元素，背篓的演绎伸展了独特的建筑形式，必将成为当地的城市商业标志。

三大片区分别为五星级酒店、大型商业综合体与高档度假社区。

The site is located on Yalong Bay in Sanya, Hainan, surrounded by around 20 five-star hotels.

Two large entrances are located in the North and South end of the site, separated by Yefeng Avenue. The hotel and recreational area are located in the northern part of the site, and have a strong connection enabling them to cater to the collective needs of the visitors.

The traditional elements from the location are introduced into the design of the recreational area. The idea of a woven basket also influenced the architectural form and will become the city's commercial landmark.

Three areas that the complex is composed of are: five-stars Hotel, a large commercial complex, and upscale resort community.

总用地面积：	6.76 ha.
总建筑面积：	93 506 m²
容积率：	0.84

Site area:	6.76 ha.
Gross floor area:	93,506 m²
F.A.R:	0.84

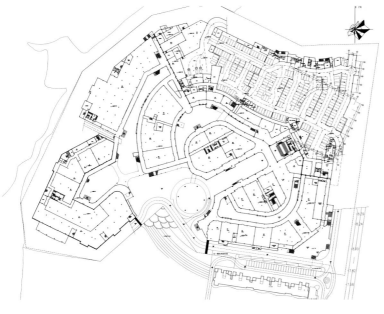

地下一层平面图
Basement 1st Floor Plan

一层平面图
1st Floor Plan

二层平面图
2nd Floor Plan

Client: Zhuhai Kingsoft Corporation Limited
设计时间 Design: 2013

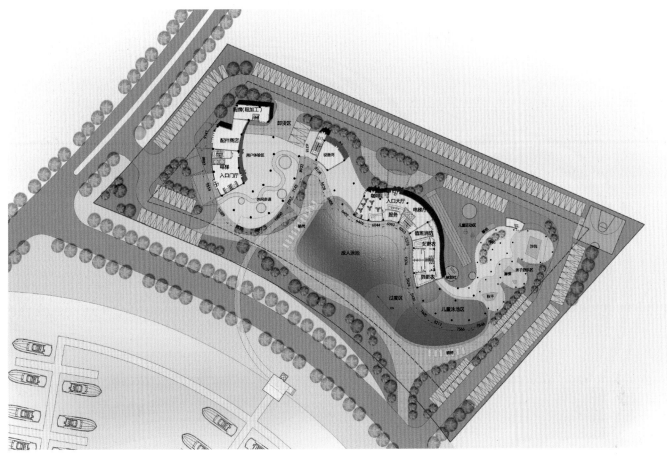

总平面图 Site Plan

淇澳游艇会位于珠海市淇澳岛南芒果湾北侧,包括陆地会所和海上游艇停泊区域两部分,拟建造一站式主题休闲度假俱乐部。

俱乐部以及贵宾会所的设计吸纳了众多的航海元素以并展现航海和水上运动的精神"自由、自然、挑战"。内部空间的设计以舒适体验为核心,功能区的布局在充分考虑海面景观最大化的前提下进行,朝海延展面的最大化保证了公共空间,例如,办公、餐饮、水疗及泳池的景观优势。所有的贵宾客房均有优等海景。

The site is located in the northern part of Nanmangguo bay, Qi'ao of Zhuhai city. The project provides both club on the ground and yachts landing area to create a holiday themed yacht club.

Club design absorbs navigation element to promote the "free, nature and challenge" spirit. The inner space design mainly focuses on luxury and comfort experience of customers. The priority of the layout is to maximize the magnificent ocean view, therefore, a curved facade along waterfront is designed to accomodate public space such as office, restaurants, spa, and swimming pool, at the same time providing ocean view to every VIP room.

总用地面积:	1.4 ha.
总建筑面积:	14 820 m²
容积率:	1.0

Site area:	1.4 ha.
Gross floor area:	14,820 m²
F.A.R:	1.0

多方案选择

二层平面图
2nd Floor Plan

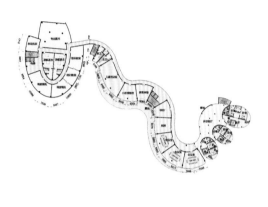

三层平面图
3rd Floor Plan

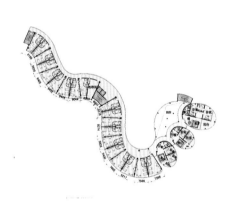

四层平面图
4th Floor Plan

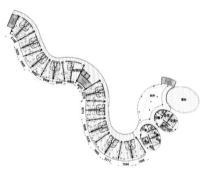

五层平面图
5th Floor Plan

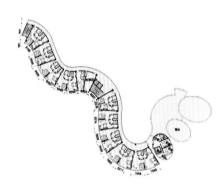

六层平面图
6th Floor Plan

哈尔滨爱建滨江国际交银大厦
Communication Bank Tower, Harbin

中国，哈尔滨 Harbin, China
业主：上海爱建股份/哈尔滨爱达投资
Client: Shanghai Aijian/ Harbin Aida Investment
设计时间 Design: 2003
建成时间 Completion: 2007

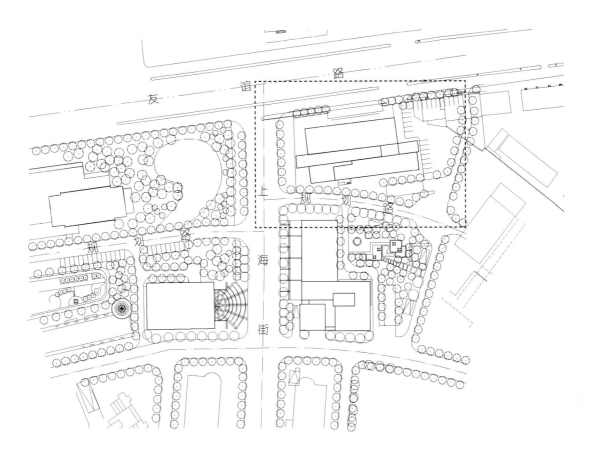

总平面图　Site Plan

哈尔滨交银大厦位于哈尔滨友谊路上，作为KFS设计的哈尔滨爱建滨江国际的一部分，它是交通银行以办公为主的综合性大厦，可以远眺松花江。设计的目标是在满足功能需求的同时，营造一幢既与冰城文脉相协调，又不失信息时代独特品格的建筑体。建筑西侧宛如松花江畔一挂瀑布的玻璃迭落体为设计的重点，亦为信息时代品格的象征，构成符合业主精神特质的设计主题。绿化、树木、植被、跌水和瀑布等自然要素通过架空、渗透等各种手段，使环境延伸入建筑，又使建筑融入城市环境，一起共生共长。办公人员紧张之余可在此小憩、沟通、眺望松花江，来此商业洽谈的人们可以在此等候、思考、交谈，共同被唤起愉悦的情绪。设计用现代的材料和独特的手段演绎建筑的宁静、优雅，成为哈尔滨有口皆碑的最独特的建筑之一。

The site located in downtown Harbin at the "T-cross" intersection of Youyi Road and Shanghai Street. The site has a direct view to the Songhuajiang River to the north. The design is based on the interior plan expression on the exterior of the building. The local culture of Harbin has a strong influence on the design for the exterior elevation. The architectural detail of the elevation is conservative yet imbued with a subtle elegance. The exterior stonework is consistent with the Harbin architectural heritage and fulfils the modern building expectations of Harbin. The "ice-city" of northern China is reflected in the cascading exterior glazed design of the west elevation with a frozen waterfall image to the urban cityscape. The facade opens up the interior of the building as a significant gesture symbolizing the connection of the city and the information technology era.

用地面积：	0.55 ha.
总建筑面积：	25 000 m²
容积率：	4.6

Site area:	0.55 ha.
Gross floor area:	25,000 m²
F.A.R:	4.6

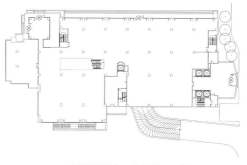

一层平面图　1st Floor Plan

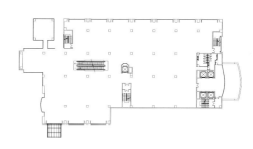

二层平面图　2nd Floor Plan

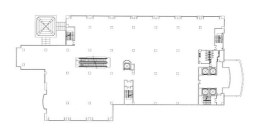

三层平面图　3rd Floor Plan

标准层平面图　Typical Floor Plan

上海长宁舜元大厦(北大青鸟)
Shunyuan Office Tower (Beida Qingniao), Shanghai

中国,上海 Shanghai, China
业主:上海北大青鸟企业发展有限公司
Client: Beida Qingniao Enterprise Development Inc.
设计时间 Design: 2002
建成时间 Completion: 2008

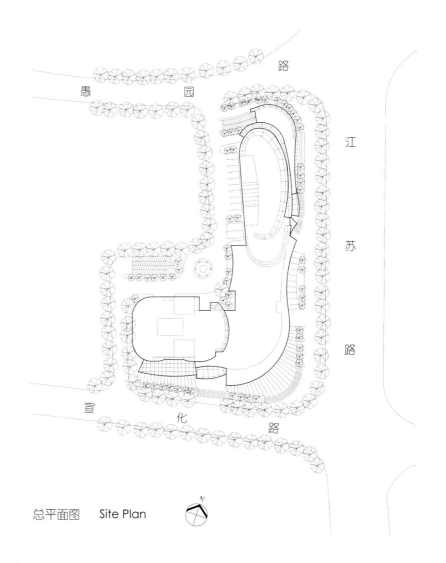

总平面图　Site Plan

上海舜元大厦位于上海长宁区江苏路、愚园路与宣化路交汇处，设计有办公、购物、居住等功能，并创造舒适、优美、便捷、高雅的环境氛围，增强该地区的场所特色。

为了保持及强化沿江苏路上的空间体量效果，将办公楼沿江苏路展开，与路对面的办公大厦相呼应，形成办公楼群以获得开阔的视野；为了创造适宜居住和办公的条件，将公寓及办公功能沿宣化路布置，既取得好的日照，又避开城市主干道的喧嚣；为了在有限的地块上合理地处理交通，把车行出入口分设在愚园路及宣化路，自然地解决两个主体建筑的出入口问题，而沿江苏路的商业裙房也获得了最大展开面，其与步行空间内部自然围合成一个尺度宜人的内庭院，使其能够容纳各种功能要求，提高基地的灵活使用程度。

办公楼结合地形采用了扁心筒布置，使办公面积既经济有效，又不失外形的活泼流畅，保证了一个挺拔优美的外观。公寓采用单元平面拼接式，优美的轮廓与办公楼相得益彰。裙楼三层以商业为主，建筑形象呈现出曲形流线，在周围平实的建筑风格中，别具风格。通透的玻璃、光滑的铝板、细致的不锈钢饰使整个群体一气呵成、流畅优美。

用地面积：	0.7 ha.
总建筑面积：	45 000 m²
容积率：	6.3

The building is in Shanghai's Changning District. Situated in the southwestern corner of the junction of Jiangsu Road and Yuyuan Road, it is close to Jiangsu Road, to the east, and Jiangcheng Road, to the west. The site extends to Xuanhua Road, to the South, and Yuyuan Road, to the north. The project is a development of luxury mixed-use buildings that include a luxury hotel, serviced apartments, professional office space, and commercial facilities.

Style and simplicity are the main design criteria for the development of the high profile landmark in the community. Both criteria find their expression in the facade through the contrast and consistency detailing. The stereotypical flat, monotonous and continuous facade has been consciously avoided. They under-line the use of colour and intelligent detailing. This clean architectural language brings consistency to the designs of the serviced apartments and the high-end office towers. The commercial facilities form the commercial centre of the development. The development is enriched through the use of artificial hi-tech computer systems, green facilities and a generous amount of parking spaces.

Site area:	0.7 ha.
Gross floor area:	45,000 m²
F.A.R:	6.3

上海浦东世纪大道长泰国际金融大厦
Changtai Office Tower, Pudong, Shanghai

中国，上海 Shanghai, China
业主：长甲集团
Client: Changjia Group
设计时间 Design: 2002
建成时间 Completion: 2008

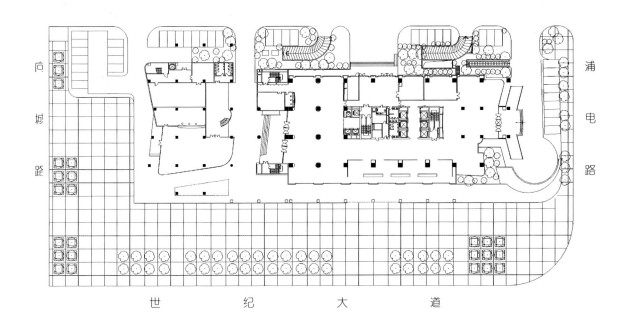

总平面图　Site Plan

上海长泰国际金融大厦位于上海浦东新区世纪大道上，北面为期货交易大厦，南临世纪大道，是一座以办公为主体，兼具会议、娱乐、银行、保险等用途的综合性大厦。办公楼主体及裙房最大限度地沿世纪大道展开布置，具有开阔的视野和良好的世纪大道景观。主要车行出入口设于东侧的浦电路上，办公主入口设于东面，其他的服务性出入口均设于基地北面，使基地南面得以留出10 m宽的绿化带。

为了在狭窄的基地上创造一个优雅舒适的外部环境，建筑底部被设计成开放式的室外广场，面向与之相邻的公共绿地，并向世纪大道敞开，同时结合世纪大道已有的景观绿化，设置了室外休息广场及绿地、树木、雕塑等小品，使整个基地与世纪大道景观相融合，达到了应势借景的效果。在主体办公楼内部，每隔四层均设有一个通高的空中花园，透过它可欣赏到世纪大道壮丽的景色。空中花园加强了办公楼内不同楼层的联系，为员工在工作之余提供了一个宁静优雅、舒展身心的地方，也提升了办公楼的内部环境质量。

Located in the Century Avenue in Pudong, the building is a mixed-use office complex surrounded by contemporary high-rise buildings. A height restriction of 100 m generated a "neo-classic" design. Stone and glass detailing is used on the facade to make a "distinctive architectural design statement". The building exterior presents a solid and steady luxury quality that reflects the client's status in the business community. The development contains commercial, entertainment, assembly hall, and institutional uses. The design concept is the creation of an elegant architectural building statement that compliments the neighbouring cityscape.

An exterior pedestrian plaza opens toward Century Avenue adjacent to the public green space. A curved glass atrium forms the pedestrian west entrance to the building and the main entrance to the offices faces the future trade centre on the east. A portico feature on the ground floor directs people into the indoor space and forms an active living space with an outdoor leisure plaza, greenery, trees, and sculpture features. A high quality indoor environment offers a natural living style enhanced by a sky garden that connects the different storeys.

用地面积:	0.9 ha.
建筑面积:	57 000 m²
容积率:	5.0

Site area:	0.9 ha.
Gross floor area:	57,000 m²
F.A.R :	5.0

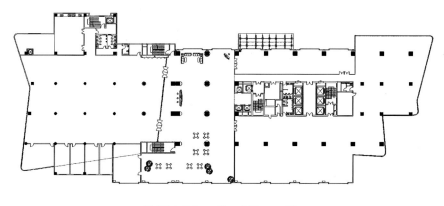

二层平面图   2nd Floor Plan

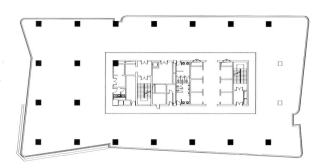

标准层平面图   Typical Floor Plan

上海浦东东晶国际办公大厦
Dongjing International Office Complex, Pudong, Shanghai

中国，上海 Shanghai, China
业主：上海东道置业有限公司
Client: Shanghai Dongdao Real Estate Inc.
设计时间 Design: 2002
建成时间 Completion: 2005

总平面图　Site Plan

东晶国际办公大楼位于浦东大道以南、源深路以东，其中浦东大道为东西走向交通大动脉。地块呈狭长形东西走向，沿地块西南源深路为有"小外滩"之称的浦东特色办公建筑，地块西有城市道路绿化带，南侧"上海滩花园洋房"为多层低密度住宅，地块东为相同用地性质地块小区。基地用地面积为1.6 ha.，总建筑面积66 000 m²，其中办公楼总高24层（不含屋顶设备层），建筑面积24 000 m²。

本案意在从建筑文化和环境人文角度着手，用崭新的时代建筑语言，建设一个环境优美、设施完备、风格独特的国际化办公区。多种功能既合又分的布局，实现多赢格局。

100 m高的东晶国际办公大楼以其独特、优雅的形象，从附近的办公建筑中脱颖而出。夜晚璀璨的灯光，沿街的"火树银花"为城市增光添彩，使人沉浸在时尚、高雅的氛围中。办公楼平面为一个方形塔楼，标准层建筑面积1 000 m²左右，适应自由灵活的分隔，每层分别设置空调机组，满足小型办公单位的需要。

用地面积：	1.6 ha.
总建筑面积：	66 000 m²
容积率：	4.0
办公楼建筑面积：	24 000 m²

The project is situated at Pudong Avenue, a major traffic path from east to west, east of Yuanshen Road. The site is a narrow rectangle from east to west. The area is focal point of the Pudong's office parks. There are city green belts west of the site. Shanghaitan Garden House is located in the south of the green belt. The site area is 1.6 ha. The gross floor area of the development is 66,000 m². This includes a 24-storey office block with a total GFA of 24,000 m².

The design development program of the office space has been judiciously refined in order to encourage the mutual benefit of the different functions. A new architectural style introduces a design image of an international multi-functional complex within a relaxed environment.

A 100 m high landmark is an architectural statement that in itself is a unique and elegant image. Ornate neon lights at night dazzle the night sky with the building seemingly garbed in a sparkling trendy outfit. The office floor plan area, of approximate 1,000 m², is designed with an open floor plan to facilitate an easy subdivision into smaller individual areas.

Site area:	1.6 ha.
Gross floor area:	66,000 m²
F.A.R:	4.0
G.F.A of office:	24,000 m²

中国，哈尔滨 Harbin, China
业主：上海爱建股份/哈尔滨爱达投资
Client: Shanghai Aijian/ Harbin Aida Investment
设计时间 Design: 2002
建成时间 Completion: 2008

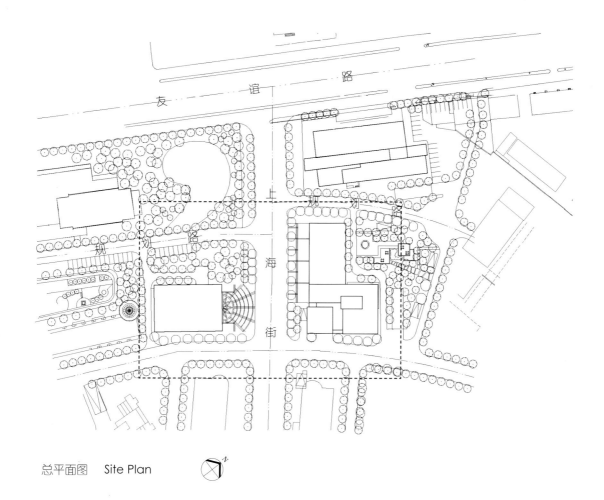

总平面图　Site Plan

整个爱建滨江国际社区以中心广场为核心，呈环状布置，本基地属于其西北片区。本项目为两幢20层高的居住办公一体建筑，坐落在友谊路与上海街交汇处。项目用地为0.9 ha.，总建筑面积50 000 m²。

两栋居住办公一体建筑隔着上海街两两相望，形成哈尔滨爱建滨江国际社区的门户建筑和标志性形象。

两栋居住办公一体建筑主要采用大玻璃及铝板材质，与沿街已建及保留建筑呼应，挺拔简练的体型塑造手法使其成为本街区的新地标。考虑当地特点和功能性质，色彩上更多运用中等明度及中性色调，整体形象既统一又显俏丽，让人们四季生活在生动活跃的氛围中。

用地面积：	0.8 ha.
总建筑面积：	50 000 m²
容积率：	5.8

This project includes two SOHO buildings, on each side of Shanghai Street. They are located at the intersection of Youyi Road and Shanghai Street, painting an icon image in Aijian New Town.

The facades of the two SOHO Buildings, covered in aluminum and glass cladding, reflect the existing architecture along the street. The architectural language of clean and straight shapes has made them the new landmark of the local area. Taking the characteristics and functional properties of the local area into consideration, KFS adopted bright colors with a neutral tone. The whole image becomes a playful addition that unifies and creates a vivid living and vibrant atmosphere.

Site area:	0.8 ha.
Gross floor area:	50,000 m²
F.A.R:	5.8

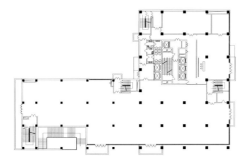

A栋1层平面图　A#　1st Floor Plan

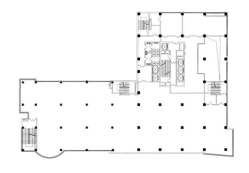

A栋2层平面图　A#　2nd Floor Plan

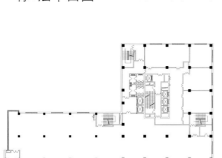

A栋3层平面图　A#　3rd Floor Plan

A栋标准层平面图　A#　Typical Floor Plan

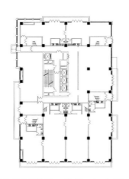

B栋1层平面图　B#　1st Floor Plan

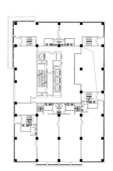

B栋2层平面图　B#　2nd Floor Plan

B栋3层平面图　B#　3rd Floor Plan

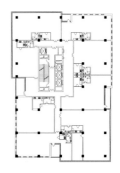

B栋标准层平面图　B#　Typical Floor Plan

上海加拿大梦加园
Dream Home Canada, Shanghai

中国,上海 Shanghai, China
业主:加拿大林业创新投资公司
Client: Canada Forestry Innovation Investment Company
设计时间 Design: 2003
建成时间 Completion: 2006

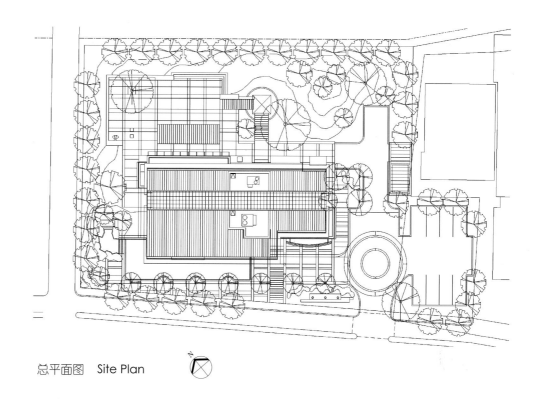

总平面图　Site Plan

上海加拿大梦加园是加拿大卑斯省政府所辖的林业创新投资公司在上海的一项工程，旨在推广卑斯省森林资产的应用。该项目将成为一个木结构建筑在中国建筑行业中的示范和样板工程。

梦加园作为一个示范工程，除了传统的产品展厅形式外，建筑的本身即展示了木结构的产品和技术在别墅式住宅和多户组合住宅中的运用，包括内、外部的木装修材料以及非住宅产品的内容。在整体布局上，建筑布置于近街的一侧，这样在基地沿河一侧空出大片花园绿地以作室外平台、绿化及展示花园木家具之用。停车位及卸货位位于基地的南侧。

在这个项目上，设计师希望不仅提供一个展示的场所，更希望设计本身从内到外向中国的工程师、建筑师及开发商展示木结构建筑、大型木材的施工技术设计及加拿大的建筑材料。从而在木材建筑及其他建筑中推广加拿大卑斯省木材及配件，同时，对中国设计师、现场监理、木结构指导师进行木结构建筑的培训和交流，以此促进该系统在中国的发展。

The Dream Home China Project (DHCP), is associated with the Forestry Innovation Investment Company, that manages and supports programs and design initiatives to increase the value of the forestry assets, on behalf of the Government of British Columbia. The DHCP project proposes introduction to the Chinese construction industry of the construction of a wood-frame demonstration project in a high-profile location in a major Chinese city.

The three-year plan is a programmed comprehensive approach to the promotion of BC building products and wood-frame construction techniques. The demonstration project includes types of wood construction in single and multi-family residential structures, interior and exterior wood finished products, and some non-residential construction applications. The centerpiece of the plan is a 894 m^2 multi-purpose presentation centre.

The centre highlights advanced wood technologies and heavy timber wood products in its construction, and includes space for demonstrations of BC commodity and value-added forest products, offices, space for receptions, and training and related commercial services. The site is fully landscaped, and profiles suitable BC wood products.

用地面积：	0.3 ha.
总建筑面积：	1 500 m^2
建筑密度：	15%

Site area:	0.3 ha.
Gross floor area:	1,500 m^2
Building density:	15%

上海海纳科技研发大楼
Haina Hi-Tech Building, Shanghai

中国，上海 Shanghai, China
业主：上海大华集团
Client: Shanghai Dahua Group
设计时间 Design: 2002
建成时间 Completion: 2005

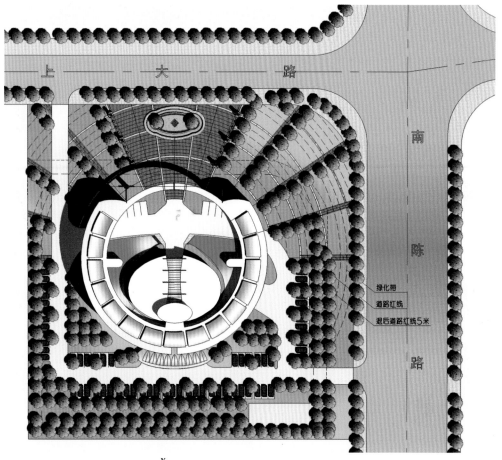

总平面图　Site Plan

"上海大学科技园四通纳米港"高新科技园区,是为纳米科技提供的一个研发基地。上海海纳科技研发大楼作为四通纳米港中的标志性建筑,业主希望建筑体能够成为科技进步与社会发展的一个象征。

建筑,作为一个载体,如何承载象征的意义便成为此项设计的基本任务。

在功能组织和体量推敲后,确定了现在的设计方案。6层高的圆形围合一个中庭和一个贝壳形的会展商务中心。敞开的中庭、对称的布局、围合的空间,用这样的建筑形式来体现"海纳百川"的意味,为办公楼确定基本建筑形态。

建筑中央安排一个大型空间的功能集合体,它本身的形象就像一颗孕育着明珠的贝壳,暗喻纳米港区对科技的孵化作用。环形的办公部分围合着一个敞开的中庭空间,支撑空中走廊的两大立柱也由此形成具科技意味的入口形式,成为纳米港区的一个象征性的门户,同时也使建筑的整体外观更为壮观。南广场地坪标高作了一个提升,以使主要入口更为注目。在有限的基地范围内,设计有较多的绿地及休闲空间。

Located at the Shanghai University Science Park of Sitong Nami Harbour, a significant "waterfront" building, the building is intended to be a strong symbol of international technological development in China. Nami's high technology reflects the rapid growth in the Chinese economy and is the guideline for the design concept that is based on the Chinese meaning of 'comprehensive'.

The entrance and introduction to the building is defined by a series of corridor columns and creates an interesting architectural statement on the building facade. The Nami entrance is elevated to heighten the building significance.

A 6-storey circular space forms the atrium in a shell-like shape that defines the commercial exhibition centre. The expansive atrium is symmetrical in layout design and is the architectural vocabulary for exhibiting the comprehensive and progressive spirit of Nami technology. The large scale of the multi-functional complex, in the shape of a shimmering shell is at the centre of the building. A circular office building surrounds the atrium.

State–of-the-art material, advanced structural design, and attention to unique detailing mirror the characteristics of the high technology industry. The landscape greenery and exterior leisure space offer a soft and subtle contrast to the building's "edge" qualities.

用地面积:	1.1 ha.
建筑面积:	8 000 m²
容积率:	0.7

Site area:	1.1 ha.
Gross floor area:	8,000 m²
F.A.R :	0.7

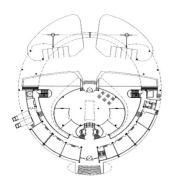

底层平面图
Ground Floor Plan

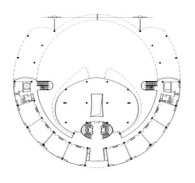

2层平面图
2nd Floor Plan

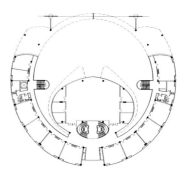

3、4层平面图
3rd & 4th Floor Plan

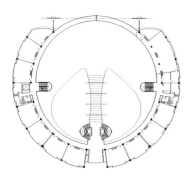

5层平面图
5th Floor Plan

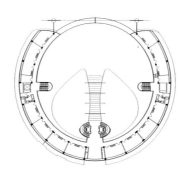

顶层平面图
Top Floor Plan

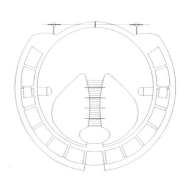

屋顶平面图
Roof Plan

派诺珠海科技园
Pilot High-Tech Park, Zhuhai

中国，珠海 Zhuhai, China
业主：珠海派诺科技股份有限公司
Client: Zhuhai Pilot Technology Co.,Ltd.
设计时间 Design: 2014

基地位于珠海市香洲区科技创新海岸南围片区科技九路北侧、创新八路东侧，周围都为已建成的工业用房，基地内地势平坦，形状为矩形。

为创造出低造价，小而美的绿色建筑，坚持三大设计原则，分别为：

1。以节俭为设计策略
2。以常识为设计基点
3。以适宜技术为设计手段

设计从理性的场地分析，建筑体量布置，形体变化，到内部空间的塑造，再到外部空间的融合，一气呵成，造就出形态优美，环境交融的感性设计。

基地出入口位于西侧，紧邻创新八路。南北侧1#楼与2#楼之间为景观水池，地面一层部分主要为架空层，屋顶为层层跌落的绿色种植屋面。运用绿色节能设计理念，从风向控制建筑走向，以最小形体系数控制整个建筑的形体。

在场地中央设置两个大型景观水池，入口广场、景观水池、步道、首层架空区域景观及场地内其他景观有机结合。

总用地面积约：	1.2 ha.
地上建筑面积：	21 600 m²
容积率：	1.8

The site is located in the south margin zone of the Innovation of Science and Technology Park in Xiangzhou district, Zhuhai city. The site terrain is flat, with rectangular shape and surrounded by industrial building.

The concept starts from rational analysis of the site, building scale and space interaction.

The main entrance is from west which next to the Xinba road. A landscape pond is located between the two main buildings. The design has elevated ground floor and stepped roof garden. The design fully consider energy efficiency and sustainability, three buildings in both north and south of the site are designed according to wind direction so that the buildings will have minimum wind impact.

Two large scale ponds are arranged on the center of the site linked with the entrance plaza, pedestrian street and elevated ground to provide a joyful landscape.

Site area:	1.2 ha.
Total Building area:	21,600 m²
F.A.R:	1.8

KFS国际建筑师事务所上海办公楼
KFS Office Building, Shanghai

中国，上海 Shanghai, China
业主：加拿大KFS国际建筑师事务所
Client: K.F.STONE Design International Inc. Canada
设计时间 Design: 2005
建成时间 Completion: 2008

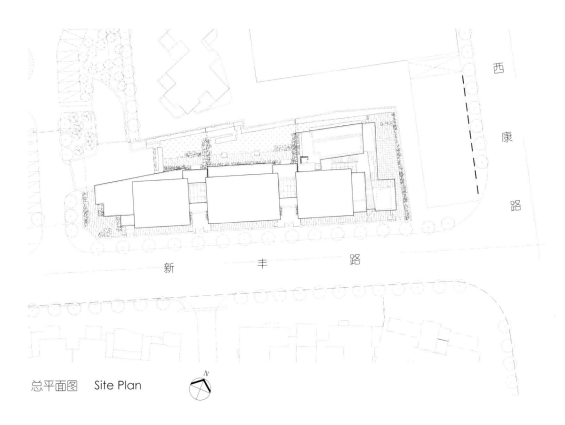

总平面图　Site Plan

该建筑包含三个独立式办公单元，除拥有各自的出入口门厅、沿街面和地下车库外，各单元之间还通过中庭连接成一体，可分可合的总体布局为最终的使用带来极大的灵活性和方便性。该建筑地上三层，建筑周边设有精致的景观园林，立面形式采用红砖外墙和现代的玻璃幕墙相结合的手法，利用局部退台形成高低有致的外观形态，创造出一种全新的小型办公楼设计模式。

用地面积：　　　　　　　0.2 ha.
总建筑面积：　　　　　　3 000 m²

The building is composed of three individual office units with separate independent entrances and exits that face the street with underground parking. The three buildings are each separated by atriums that perform as transparent 3-storey glazed linkages. This offers great flexibility for the final functional layout. These 3-storey office units are accompanied with lobbies and soft elegant landscapes. The facade designed in a Neo Classic style of natural brick masonry and curtain wall, to create new mode of small office design. The sloped metal roof appears to float above the glazed exterior with steel support structural members that offer a delicate structural contrast to the solid masonry.

Site area:　　　　　　　0.2 ha.
Gross floor area:　　　　3,000 m²

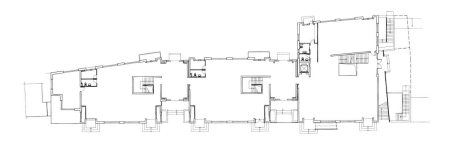

底层平面图　1st Floor Plan

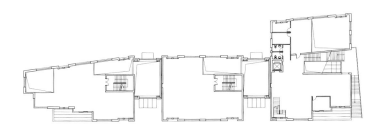

二层平面图　2nd Floor Plan

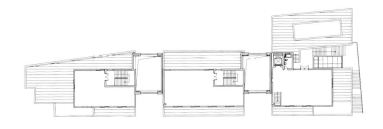

三层平面图　3rd Floor Plan

立面图　Elevation

上海苏河一号（华森钻石广场）
1# Suzhou Creek (Huasen Diamond Plaza), Shanghai

中国，上海 Shanghai, China
业主：上海丹林房地产开发有限公司
Client: Shanghai Danlin Real Estate Co.,Ltd.
设计时间 Design: 2005
建成时间 Completion: 2010

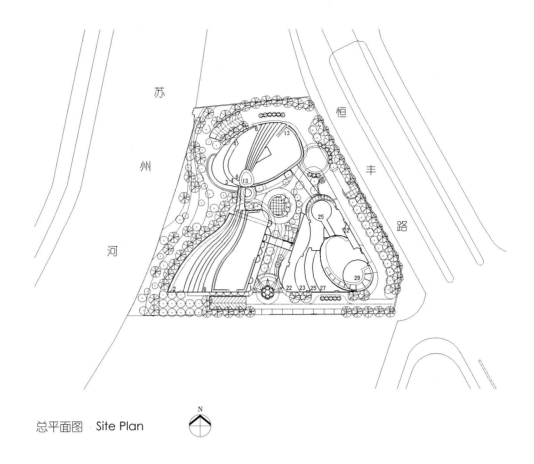

总平面图 · Site Plan

上海苏河一号（恒丰路华森钻石广场）位于上海市中心核心区内，北临苏州河。本设计从建筑内部空间的有效利用到苏州河环境与建筑艺术的展示，力求创造一组全方位多功能的综合社区，以达到环境与建筑的融合共存。

沿苏州河边有45度的高度限制，因此自然而然将三幢楼由低至高、由西向东阶梯状布置，其中两幢沿河南北展开呈流线形布置，以获得最大的景观展开面。平面流线层层跌落，形似"如意"，轻盈流畅，不仅意形并济，同时在空间体量上可视为办公主楼的裙房，以减少高层建筑对河道及人的压抑感。

东侧办公楼和西侧公寓式酒店及办公综合体之间布置了一条商业街，既可以作为商业用途，又兼有休闲和交通功能，紧急时可用做消防通道，同时在空间上有重要的衔接和过渡作用。另外通过一些中庭及天桥连廊将建筑连成一线，实现了立体的空间组织，形成一个集商业、办公、SOHO（居住办公一体建筑）、休闲、娱乐、景观于一体的"24小时高档国际商务社区"。

The Suzhou Creek #1 is in the centre of Shanghai and a little north of Suzhou Creek. The architectural design creates a multi-functional development that harmoniously co-exists with the environment through the efficient utilization of the interior architectural exhibition space and the functioning of the Suzhou Creek as natural environment.

The design of the three buildings is a step-up from west to east following the 45-degree zoning development control line for new buildings adjacent to the Suzhou Creek. Two buildings are arranged in the north-south corner for the maximum exposure to the Suzhou Creek. Streamlined built volumes rise fluidly up one after another like clouds in the sky, mitigating the oppressive nature of the high-rise buildings to the creek and open spaces.

A commercial street, between the east office building and the hotel and office building in the west, serve as a transportation corridor, fire-route and recreational space. Atriums and a pedestrian bridge connect the buildings and offer a variety of spatial scale to the different spaces. The retail, offices, small office homes, recreation and landscape generate an active 24-hour international business community.

用地面积：	1.4 ha.
总建筑面积：	68 000 m²
容积率：	5.0

Site area:	1.4 ha.
Gross floor area:	68,000 m²
F.A.R :	5.0

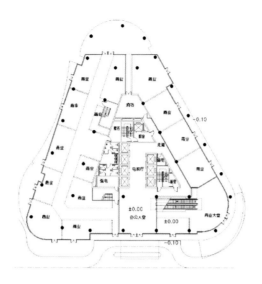
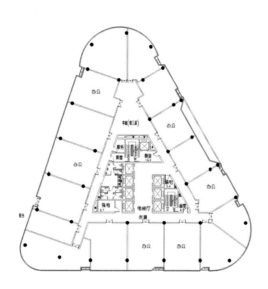
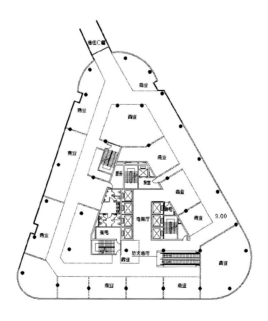

办公楼平面　Office Tower Floor Plans

上海静安达安河畔雅苑
Da'an Riverside Tower, Shanghai

中国，上海 Shanghai, China
业主：上海达安房产
Client: Shanghai Da'an Real Estate Co.
设计时间 Design: 2003
建成时间 Completion: 2006

总平面图 Site Plan

地块位于上海市静安区西苏州河边。项目意在创建一个高品质的小户型办公、商业与住宅综合体，用一种新的时代语言，建设一个环境优美、设施完备、风格独特的"国际化"商业、办公、居住建筑。

如何欣赏上海具有人文自然景观的苏州河成为设计的关键，解决沿河建筑规划控制带来的问题是设计的重点。在整体规划上，简约的总体布局，富有戏剧性的构图，使得空间层次变化多样。在西苏州河上让出更多的空间，留下更多的阳光，减轻了对城市主干道的压力，为内部空间品质奠定牢固的基础。

在住宅设计上，每户70~90 m²的房型采用南北跃层式布局的方式，让每户居民既有充足的南向日照，又能欣赏到北侧美丽的苏州河景。部分单元可以组合分拆。层层退台的体型，制造了顶层多个北向景观露台，顶部局部形成舒适的大户观景房型，同时也丰富了建筑立面造型。

The project is at the border of Suzhou Creek in Jing'an District of Shanghai. It creates a small-scale apartment and office space quarter of high quality. This quarter forms an international community with graceful spaces, natural facilities and a distinctive architectural style.

A simplified master plan forms a dramatic space with a variety of changing views. The large space is set back from Suzhou Creek allowing more sunlight into the site and provides the city street with more usable open space.

The 70~90 m² apartments are arranged in a "jump floor" system that promotes the enjoyment of the sunshine from the south and the view of Suzhou Creek from the north. Design flexibility makes it possible for some apartments to be incorporated into larger units. The "step design" of the building volume creates many roof terraces and luxury penthouses that enrich the architectural vocabulary of the building.

用地面积：	0.43 ha.
总建筑面积：	16 000 m²
容积率：	3.1

Site area:	0.43 ha.
Gross floor area:	16,000 m²
F.A.R:	3.1

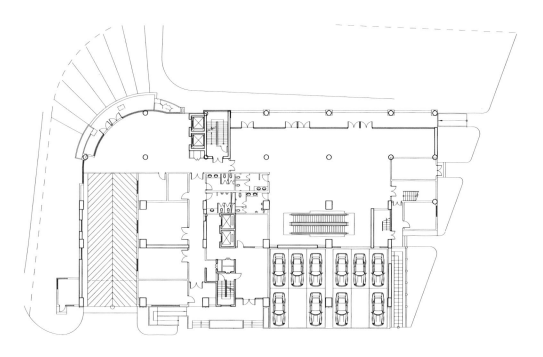

底层平面图　1st Floor Plan

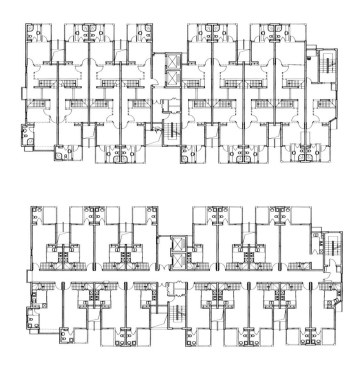

标准层平面图　Typical Floor Plan

珠海金山软件园区
Kingsoft Headquarters, Zhuhai

中国，珠海 Zhuhai, China
业主：金山软件股份有限公司
Client: Kingsoft Corporation Limited
设计时间 Design: 2009

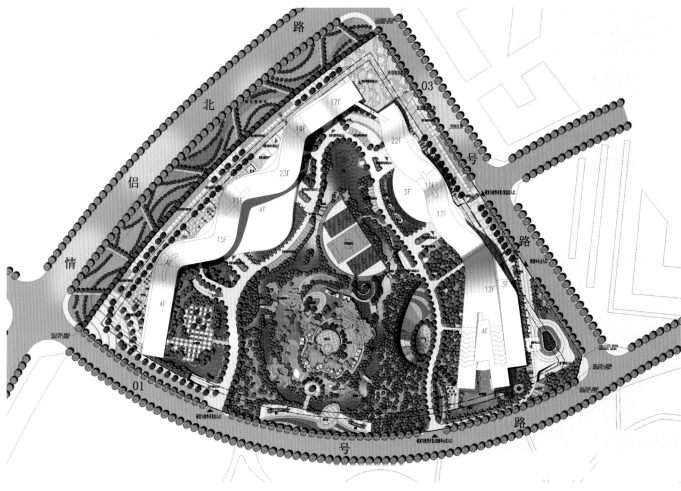

总平面图　Site Plan

本案位于珠海高新区总部基地西南侧地块，规划用地南临大海，东北依石坑山，靠山临海，景观环境优越。

建筑外形的灵感来自周围群山和南面的海浪，从而形成了三大特点：

1. 成熟的技术
通过时尚的建筑形式和简洁的结构布局方式营造最有效的功能空间，有利于节约成本、分期建设、缩短建设周期。20 m的进深尺度保证了办公、会议等主要使用房间均能自然通风。

2. 时尚的外观
"山水"元素的引入，构建了时尚的地标性建筑外观。

3. 超级海景
主要建筑物都沿路规划，以保证最大的朝海面。建筑设置似双手环抱着基地，最大限度利用天然的观景面。同时为避免出现背向海面的用户，采用了20 m的进深尺度，基本实现全面观海。

用地面积：	9.7 ha.
总建筑面积：	145 000 m²
容积率：	1.5

The development is in Zhuhai, south of the sea and northeast of Shikeng Mountain. The mountains to the north and sea waves from south, were mandatory features in the design concept. The drive from downtown toward the north, or sailing from sea, the Kingsoft Park's shape mirrors the surrounding mountains.

The primary design features three dominant conditions.

1. The established technology of modern architectural forms and minimum structure design create efficient functional space. It reduces the total costs, shortens the construction phases, and facilitates the interior plan spaces and volume spaces. Aeration of main offices and conference rooms is provided with 20m depth.

2. High Fashion facade design concept of "mountain and sea", established a landmark icon architecture. The architecture itself demonstrates the contemporary style, with the facade use of blue curtain wall, immense wave aluminum roof and balconies popping from the main facade. The building looks magnificent from every angle, especially viewed from the sea.

3. Superb sea-view of all the main buildings settled along the sea-line, extrapolating every possible view of the sea. The architecture looks like two arms encircling the entire site, each single room can properly view the sea in 20 m depth.

Total area:	9.7 ha.
Gross floor area:	145,000 m²
F.A.R:	1.5

一层平面图　1st Floor Plan

地下室平面图　Basement Floor Plan

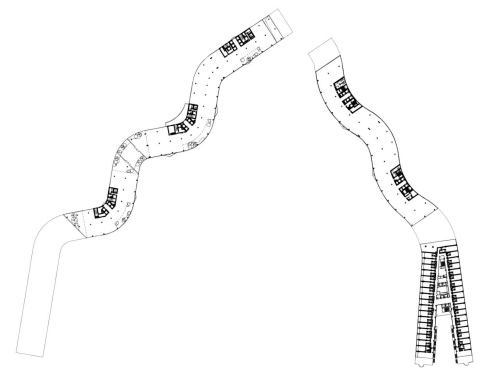

标准层平面图　Typical Floor Plan

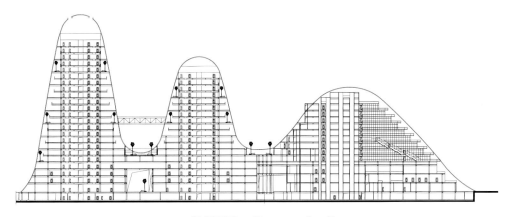

一期剖面图　Phase I Section

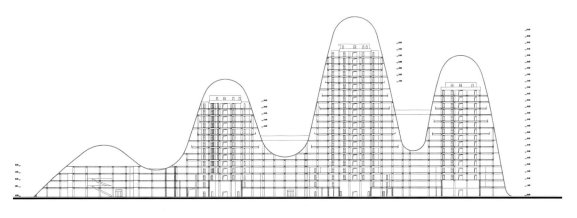

二期剖面图　Phase II Section

171

杭州西湖勾山国际
Goushan International, the West Lake, Hangzhou

中国，杭州 Hangzhou, China
业主：杭州涌金置业投资有限公司
Client: Hangzhou Yongjin Co.,Ltd.
设计时间 Design: 2009
建成时间 Completion: 2015

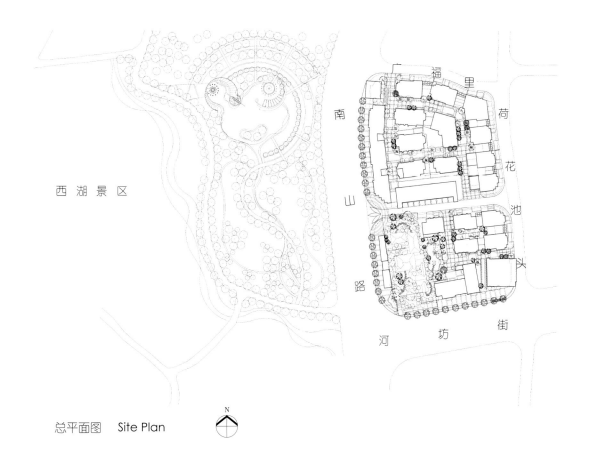

总平面图　Site Plan

杭州勾山国际项目与西湖景区零距离，与中国美术学院等杭州优秀建筑相邻。要求建设为高端企业会所及精品商业，保留基地内部的老建筑，满足规划上的15 m的限高，并于整个西湖景区的历史风貌互相呼应。

根据项目的情况所制定的设计目标和原则：
1。独栋最大化、价值最大化；
2。外部景观利用最大化、内部景观庭院化；
3。主要沿街面商业开放化、内部空间安静私密化；
4。地下空间利用最大化、交通一体化；
5。独立的出入口、风格独特的立面。

具体设计处理上：
（1）内部空间和建筑的有序设置将业主对物业的需求和城市语境的表达做了平衡处理，针对独幢利益最大化的顶端企业会所的定位，设计采用了稳重大方的体块处理，同时在外墙材质和门窗等部位做了与总体环境的协调设计。

（2）以尊重杭州传统风貌为前提，以现有周边建筑为依据，通过青砖、砂岩与玻璃等传统建筑材料，打造融入杭州，融入西湖的的建筑风格，并通过木材、红砖、轻钢等材料的点缀营造出更加细腻、别致的高档品位，力求让建筑在南山路精品建筑群中脱颖而出。

总用地面积：	1.3 ha.
地上总建筑面积：	15 100 m²
地下总建筑面积：	16 200 m²
容积率：	1.1

This Hangzhou Goushanli project is in close proximity to the famous tourist attraction of West Lake, China Academy of Art and other outstanding architecture locations. The goal of the design is to blend in prestigious offices and luxurious shops with the traditional Hangzhou urban context. The design preserves four historical buildings and respects the 15 m building height.

Design considerations included the following. A maximum of open space and property value, the use of the natural side-yards, as exterior courtyards, the promotion of the retail shops being open to the street, and the preservation of the interior space as quiet, private and secluded. An underground area as large as possible with one parking system. Each unit has an individual entrance and an elegant facade.

The site plan convey a balance of the private owner and the public city's interests. To match the high level image of the clientele, KFS used a design language for the facades with shapes characteristic of stability and sophistication. For a harmonious surrounding environment the facade materials are chosen with great care.

KFS paid respect to the traditional Hangzhou atmosphere and building fabric through the use of materials of grey brick, sandstone and glass. The composition of materials is in harmony with West Lake. For an exquisite touch and finest class the design introduced, in addition to the traditional materialization, timber, red brick and steel. It will be the landmark of Nanshan Road.

Site area:	1.3 ha.
Above ground G.F.A:	15,100 m²
Underground G.F.A:	16,200 m²
F.A.R:	1.1

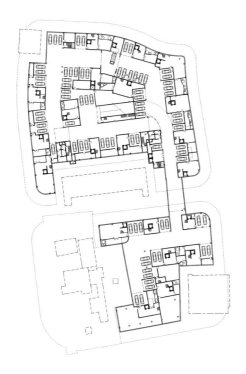

地下二层平面图
Basement 2nd Floor Plan

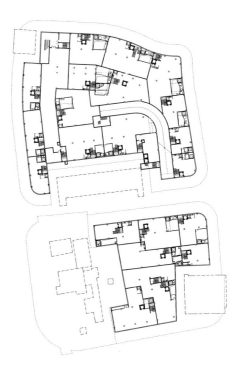

地下一层平面图
Basement 1st Floor Plan

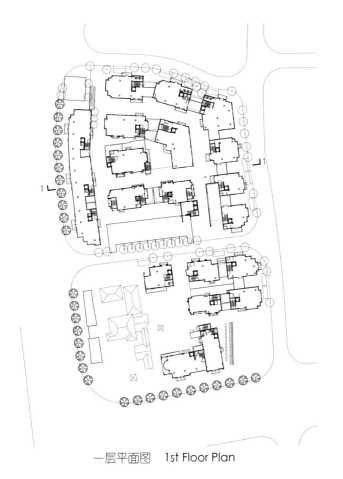

一层平面图　1st Floor Plan

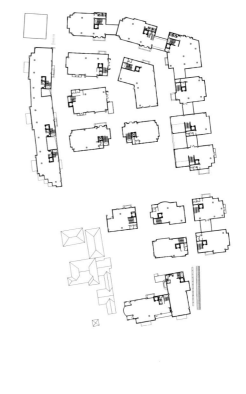

二层平面图　2nd Floor Plan

苏州唯亭君地一期
Weiting Phase 1, Suzhou

中国，苏州 Suzhou, China
业主：上海君地投资有限公司
Client: Jundi Investment Co.,Ltd.
设计时间 Design: 2010
建成时间 Completion: 2014

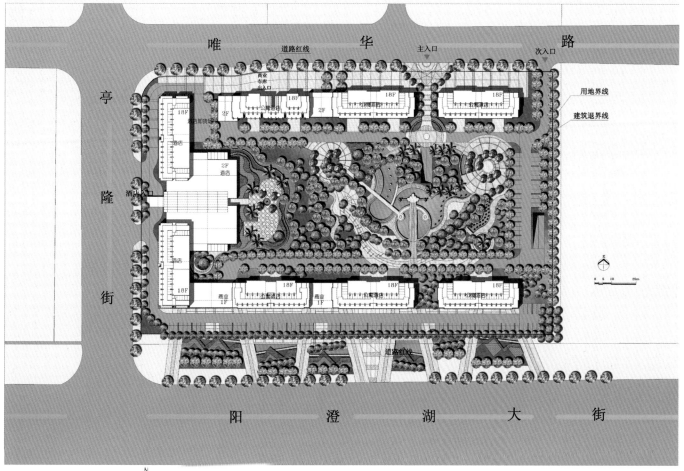

总平面图　Site Plan

本项目采用"化零为整"的设计理念，将建筑沿周边布置，在城市道路沿线形成整齐划一的城市界面，同时在内部围合出近20 000 m²的中心绿化带，成为城市的天然氧吧，让每栋房子都能一览无遗地欣赏到中央绿地的优美景色。

六栋商业服务塔楼分列绿地南北两侧。在东西向区域视线走廊上，沿亭隆街安排了一栋商业服务塔楼，和东侧相邻地块的规划酒店遥相呼应，相得益彰。商铺在街面联成一个整体，均为一层。

中央绿地成为整个景观设计的灵魂，以大片草地和乔木为主，配以适当的广场，形成丰富多变的休闲空间。沿街商业部分以硬质铺装为主，配合休闲椅，营造休闲购物的舒适空间。

立面整体采用装饰艺术风格，强烈的竖向线条、优美的顶部收分让60 m高的建筑显得挺拔有力。设计十分注重建筑细部的刻画和材料的搭配，处处体现出高贵典雅的品质。

Using formal integration as the basic concept of this project, structures of this project are linearly placed along the street, creating an orderly and unified urban element. One commercial service building is located beside Tinglong Street positioned perpendicular to the hotel. Retail shops are aligned along the street on a single floor. The store fronts are lined with pavement to enhance the shopping experience.

Six commercial service buildings are positioned at north and south end with a scenic green belt at west and east. The building act as a border in forming a 20,000 m² of green space providing refreshing and relaxing atmosphere for the inhabitants of the building.

The central green space acts as the major landscaping feature mainly consisting of large trees placed on fields of grass. The facades are inspired by art-deco style facade, emphasizing vertical lines with volumetric shrinkage at the top, giving the 60 m building a greater sense of height. The facade detailing was carefully considered though the choice of materials, displaying luxury and elegance.

用地面积：	3.7 ha.
总建筑面积：	149 000 m²
容积率：	3.03

Site area:	3.7 ha.
Gross floor area:	149,000 m²
F.A.R:	3.03

RESIDENTIAL DESIGN

居住建筑

1# Yalong Bay, Residential Building, Sanya

中国，三亚 Sanya, China
业主：海南中亚置业有限公司
Client: hainan Shenya Group
设计时间 Design: 2013

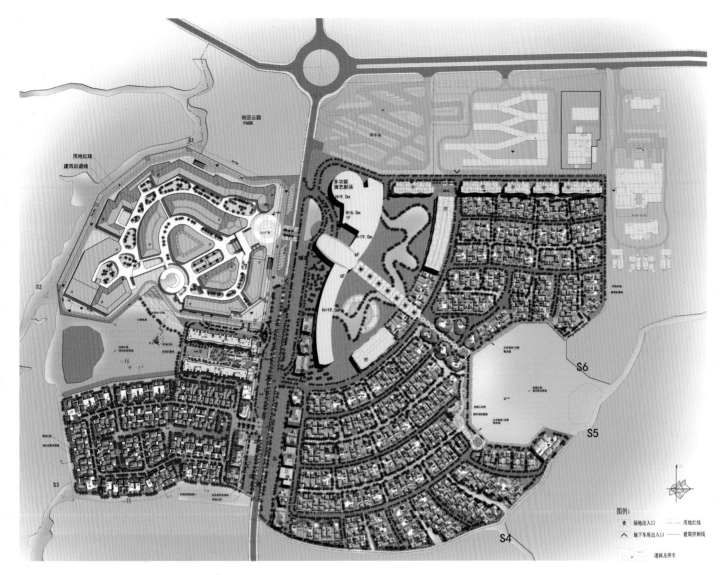

总平面图　Site Plan

海南三亚亚龙湾一号居住综合体，即AC-3地块整体呈"扇"形，且在"扇"形中部衔接规划中的城市景观湖面，"扇"形展开面长约650 m，径向约250 m。地形整体自北向南呈缓坡状，北高南低。地块西侧南部沿城市道路椰风路，西侧北部与本项目拟建设的五星级酒店用地毗邻；北侧与城市市政设施用地为邻，东侧及南侧面对规划中的运动休闲景观场地。

AC-3地块由于相对远离城市道路，拥抱内湖，隔溪流与城市休闲景观相望，适宜营造静谧的旅游度假设施。故本地块设计中，充分利用场地丰富的环境资源优势，精心营造内部水系，或汇入内湖，或流入地界周边的溪流，形成舒适惬意的内河水景空间，在山水之间形成贯通的视觉走廊，同时达到户户临水而居，将社区环境完美地与自然环境融为一体。

The radiated shape site AC-3 with arc length 650 m and radius 250 m adapts planning urban landscape lake in the middle. The overall site slope from north to south. The site is surrounded by main street Yefeng road on the south of the west part, proposing five-star hotel on the north of the west part, land used for urban municipal facility on the north, proposing sports recreation ground on the east and south.

The site which is far from the city traffic is ideal to propose a quiet holiday Resort, embracing the inner lake of the site with city view across the stream. Therefore, the landscape design aims to create a pleasant living environment with internal water system which connects either to the inner lake or the surrounding streams. The community is perfectly integrated with the natural environment with "living by the water" lifestyle.

总用地面积：	16.3 ha.
总建筑面积：	132 004 m²
容积率：	0.8

Site area:	16.3 ha.
Gross floor area:	132,004 m²
F.A.R:	0.8

上海松江安贝尔花园
The Albert Garden, Songjiang, Shanghai

中国，上海 Shanghai, China
业主：旭辉集团
Client: CIFI Group
设计时间 Design: 2014

总平面图 Site Plan

基地位于上海松江区车墩镇，紧邻车墩景视基地。地块东至影城路，南至影车路，西至车亭公路，距G15沈海高速约80 m，北至影振路。

本项目的开发定位为中密度居住社区，充分利用基地周边的区域条件，创造具有独立个性的居住空间，提供多元化的居住产品。

小区总体规划采用行列式布局，确保每栋建筑都能获得最佳朝向。考虑西侧G15沈海高速的环境不利影响，在小区的北侧与西侧布置多层住宅，东南侧布置低层联排住宅。两者之间通过中央景观带连接。规划还在小区中央打造一条南北向景观轴，营造社区运动、休闲、交流空间；并延伸至各个组团，通过景观步行道串联各个组团。在小区北侧影振路上与东侧影城路北部布置沿街商业配套用房，并在影振路主入口处形成一个30 m宽，50 m长的迎宾广场，汇聚人气并为小区居民提供一个社交的场所。

总用地面积约：	8.0 ha.
总建筑面积：	129 000 m²
容积率：	1.30

The site is located in Songjiang District Chedun Town, next to the Chedun movie studio. The site is bounded by Yingcheng Road on the east, Yingche road on the south, Cheting Highway on the west, which is 80 m from G15 Shenhai Motorway, and north to Yingzhen Road.

The Project proposes a mid-density residential area. The design utilizes the surrounding resources to provide independent and characteristic living space with a wide range of housing type choices.

The overall layout insures that each building has the best orientation, while at the same time minimizing the noises from the G15 Shenhai Motorway. In order to implement this, the plan has multi-storey apartments on the north and west side and low rise town houses on the east with a central landscape to create a connection. The north-south central landscape corridor provides spaces for interactive activities, leisure and communication. It extends to various landscape groups. The commercial buildings along Yingzhen Road and Yingcheng Road forms a 30 m by 50 m square with a main entrance on the Yingzhen Road to provide open space.

site area:	8.0 ha.
Total Building area:	129,000 m²
F.A.R:	1.30

南京秦淮河G65项目
G65 Qinhuai River Project, Nanjing

中国，南京 Nanjing, China
业主：南京鑫磊房地产开发有限公司
Client: Nanjing Xinlei Real Estate Development Co.,Ltd.
设计时间 Design: 2014

总平面图 Site Plan

规划用地为鼓楼区热河南路酿造厂地块，东至热河南路，南至淮滨路，西至二板桥，北至热河南路308号住宅小区。

设计充分考虑住宅和商业的分部关系，对北侧现有308小区住宅日照的影响，以及周边道路、绿化和高差对该地块的影响。设计将高层住宅尽量布置在地块南侧面向秦淮河的一侧，以减少对北侧住宅的影响，将二至三层的独栋商业布置在基地北侧。住宅与商业用围墙完全隔开，形成南北两块独立的区域，使彼此间的干扰最小，只在消防道路上做必要的连接。设计在商业独栋区域内布置一块集中绿地，结合入口广场形成商业步行街。每个商业独栋的入口均面向步行街开口。在基地西侧和南侧布置两个独立的地下车库出入口，将机动车流从基地外直接引入地下，形成地上完整的人行系统。

The site is in Rehe South Road in Gulou District, the former site of brewery. It is bounded by Rehe South Road on the east, Huaibin Road on the south, Erban Bridge on the west and No. 308 residence community on the north.

The design outcome is a careful consideration of the relation between commercial and residential space, the sunshade of apartments on the north and the surrounding environment. High rise building is placed on the south of the site facing Qinhuai river; 2 to 3 storeys commercial building is arranged on the north of the site to minimize interference. An open space is designed in the commercial area links to the plaza entrance to form a pedestrian street which branches to every commercial building. Car park is underground with entrances on both west and north sides leave pedestrian completely separated from vehicles.

总用地面积：	2.1 ha.
总建筑面积：	42 000 m²
容积率：	2.0

Site area:	2.1 ha.
Gross floor area:	42,000 m²
F.A.R:	2.0

青岛惜福镇泰晤士小镇
Thames Town, Xifu, Qingdao

中国，青岛 Qingdao, China
业主：鲁商置业股份有限公司
Client: Lushang property Co.,Ltd.
设计时间 Design: 2013

基地位于青岛市惜福镇，紧邻崂山景区，距离即墨市、城阳区中心、流亭机场均在10 km范围内。本项目总用地面积82 ha.。

城市社区提供一个居住与生活的全面解决之道，本设计试图将市政设施、社区服务及行政管理、多元化文体活动、多形态建筑产品、可持续发展的环保行为等全面引入社区，使之形成自给自足的、可持续发展的居住环境。

建筑风格力求具有创新性、独特性，塑造沉稳、高雅的建筑形象，在符合本项目建设区域城市整体要求的同时，使本项目的建筑形象能够体现出"英伦风情泰晤士小镇"的品牌理念。

This residential project is located in Xifu town, Qingdao city. It is surrounded by Laoshan scenic area, Jimo city and Chengyang center. The Liuting Airport is within the 10 km radius from this site, covering an area of 82 ha.

The location provides access to a number of useful facilities; including Qingdao University of Technology, and government offices encompassing municipal administration and community service. The situation of the residence promotes a diverse range of activities, architecture, and sustainable eco-friendly characteristics, creating self-sufficient living environment.

The strong linear facades have provided the buildings with uniqueness, steadiness, and elegance while the interior space allows a comfortable lifestyle. The combination of these elements created an image and concept of the "English Thames Town".

总用地面积：	82 ha.
总建筑面积：	1 287 000 m²
容积率：	1.54

Site area:	82 ha.
Gross floor area:	1,287,000 m²
F.A.R:	1.54

无锡日式服务公寓
Japanese Style Service Apartment, Wuxi

地点｜无锡，Wuxi, China
业主｜无锡鑫畅置业有限公司
Client: Wuxi Xinchang Real Estate Development Co.,Ltd.
设计时间｜Design: 2005
建成时间｜Completion: 2008

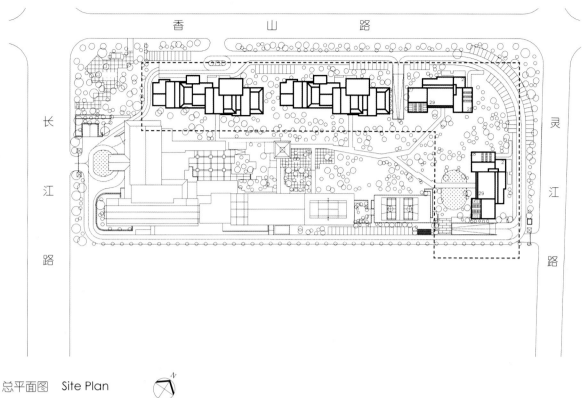

总平面图　Site Plan

无锡日式服务公寓位于无锡市无锡新区内，东侧是灵江路，西侧是长江路，北侧与香山路连接，南侧有邻接地块。用地面积约为3.8 ha.，规划地块中西北角约1 900 m²区域作为开放式城市绿地，西南角约11 900 m²区域是酒店用地，其余约25 000 m²为日式服务公寓用地，共计四栋建筑，地上建筑面积约为52 000 m²。

基地中各栋建筑（公寓、酒店等）呈回字形布置，形成了一个私密性很强的内部花园。花园约40 m宽，150 m长，精致的设计充分满足了小区用户的需要，围合性的空间强化了项目"宁静、高雅、舒适"的感觉。北侧香山路设置公寓区域的主要出入口，与酒店的主出入口互不干扰。

各栋公寓楼通过中央景观空间中的"景观回廊"连接，让业主充分享受高档社区的便利性与舒适性，享受穿行于自然景观环境中的身心满足。同时，自然地通过"景观回廊"形成各主题景观空间，体现了中式园林"移步异景"的特色。

本项目为建筑设计与室内设计的一体化设计，KFS公司完成了所有的建筑和室内设计。

The project is located in the Wuxi New Area in the city of Wuxi. The site boundaries are formed by existing roads, The site area is 3.8 ha. The 1,900 m² of land in the north-east corner of the site will be used as public urban green space. The 11,900 m² of land in the south-west corner is planned for a hotel. The remaining 25,000 m² is for Japanese style service apartments. In total the four buildings have a combined gross floor area of 52,000 m².

The architecture surrounds and shelters a 40 m wide and 150 m long inner garden. This gives the garden a private atmosphere and character. The surrounding spaces emphasize the feeling of quiet, elegance and comfort. The entrance of the apartment is located at Xiangshan Road and is separated from the hotel entrances.

The landscape corridor surrounds the centre landscape garden. This ensures that future clients will be able to enjoy the conveniences and comfort of nature. The green landscape corridor creates, a different landscape space with the characteristic of the traditional Chinese garden.

KFS completed the full design service of both exterior and interior.

用地面积：	3.8 ha.
地上建筑面积：	52 000 m²
公寓总户数：	496

Site area:	3.8 ha.
G.F.A :	52,000 m²
Total amount of units:	496

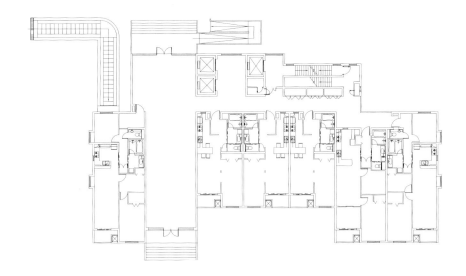

A、B栋1层平面图　A# B# 1st Floor Plan

A、B栋标准层平面图　A# B# Typical Floor Plan

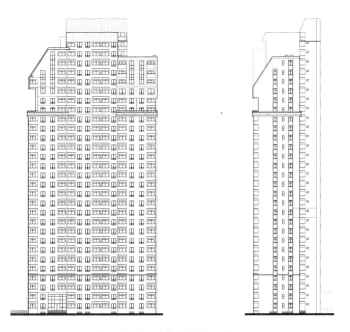

A、B栋立面图　A# B# Elevation

C、D栋1层平面图　C# D# 1st Floor Plan

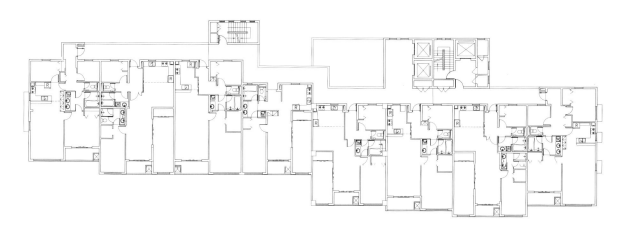

C、D栋标准层平面图　C# D# Typical Floor Plan

C、D栋立面图　C# D# Elevation

上海古北国际花园
Gubei International Garden, Shanghai

中国，上海 Shanghai, China
业主：上海新古北企业发展有限公司
Client: Shanghai New Gubei Development Go.,Ltd.
设计时间 Design: 2002
建成时间 Completion: 2006

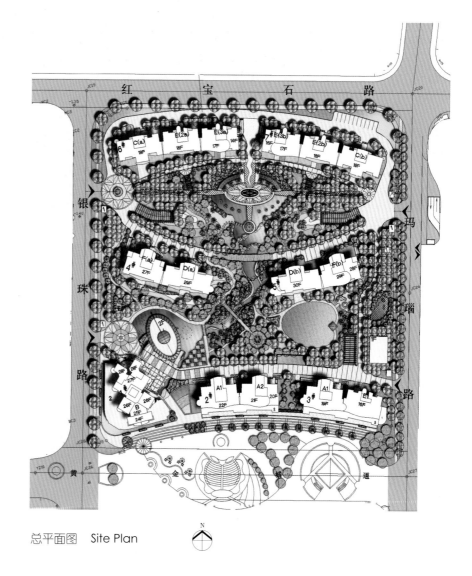

总平面图　Site Plan

上海古北新区坐落于上海中心城区西侧的黄金地段，是上海的一道亮丽的都市风景。建设中的古北国际花园住宅小区就处于这样一个城市环境中，其南侧的黄金城道是古北新区东西向重要的空间主轴，为古北新区景观中轴和公共交往空间，是一个具有商业、娱乐功能的绿化景观休闲街。基于这一规划原则，在设计中安排了前后两排建筑呈曲线形走向，中间一排建筑与后排围合成较为宁静的绿化庭院，而与前排组成具有外向、接纳特征的空间，以公寓、商业服务、会所等外向或半外向型空间与外部环境过渡、衔接。在沿黄金城道的西南角，设一步行入口广场，使黄金城道呈现出的区域文化休闲特征与社区高档的居住氛围得以相互交融，相互提升。

社区内部道路沿基地周边设置，地下车库出入口均设于小区大门附近，既避免其干扰安静舒适的居住氛围，也使得社区内部具有最大面积的步行空间。景观设计以绿化和水体相结合为主，采用局部对称和总体自由灵活的设计手法，很好地顺应了建筑本身所形成的空间特点，将建筑空间与景观设计融为一体。从会所周围流出的水面缓缓地汇成小溪，流向社区中心，而大小不一的水面、草坪、树木和铺地有机地组合在一起，带给住户优雅宁静，多样而统一的景观感受。

用地面积：	4.0 ha.
总建筑面积：	110 000 m²
容积率：	2.8

The Gubei International Garden, in the Gubei New District in the Shanghai downtown core, borders the Hongqiao development zone. The No.3 Gubei District creates a high standard in international residence development.

South of Gubei International Garden is an important east-west axis, planned as the central landscaped commercial street of the Gubei New District. The master plan significantly improves the district layout and emphasizes a human scale residential environment. The architecture style creates varied spaces of character and quality.

The curvilinear layout of the front and back building rows, detailed planning in the middle row quiet courtyards with the back row, opens the space with the front row buildings. Apartments, commercial services and clubs are transition uses to the exterior environment. The southwestern corner of the golden axis is the entrance-square that connects the community to the surrounding city neighbourhood. The recreational character of the golden axis is designed to blend into the residential community atmosphere.

Bright colored terraces and wave-like platforms endow the architecture with a free and open landscape modernity. Greenery and water combine to create a natural landscape. Water surrounds the club and forms a relaxing creek at the center of the community. Pools, grass, plants and pavement organically combine the diverse yet integrated landscaping.

Site area:	4.0 ha.
Gross floor area:	110,000 m²
F.A.R:	2.8

中国，上海 Shanghai, China
业主：上海长甲置业有限公司
Client: Shanghai Changjia Real Estate Co.,Ltd.
设计时间 Design: 2002
建成时间 Completion: 2006

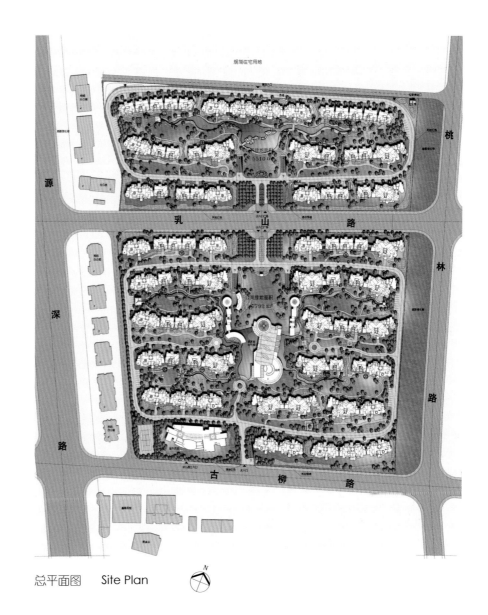

总平面图　Site Plan

上海滩花园洋房坐落在上海浦东新区陆家嘴金融贸易区内，交通便捷，生活、娱乐、商业设施完备，是陆家嘴地区唯一的低容积率的高档次住宅小区，区域条件十分优越。

在历史上，花园洋房历来是上海居住形式的最爱，居住形式的顶峰。如何在容积率为1.0，限高为18 m的地块中把五、六层的大众化建筑演化成上海的花园洋房的形式，是建筑师能否取得成功的关键。为贴合居住者亲近自然的愿望，使建筑采用了层层退台的形式，不仅使居住者感受露台的自然情趣，建筑体量更丰富，也有效地缩短了人与自然的距离，压缩了建筑的视觉体量，让建筑不再显得高耸压抑，而是体现出洋房般的亲切宜人。完全掩映在绿树花香之中，体现了都市园林的设计意境，花园洋房品质就此被塑造了出来。

用地面积：	14.4 ha.
建筑面积：	136 000 m²
容积率：	0.94
住宅总户数：	635户

Shanghaitan Garden House is a residential development in the Lujiazui Area of Shanghai with established traffic circulation advantages coupled with commercial, and entertainment amenities. This low-density luxury residential development is the single low-rise high quality development in the area.

Green space and water are strong natural features that attract people to this active and vibrant city. The creation of a green and natural living environment is the single most important design consideration. The business success is the subsequent increase in the real value of the property. Apartment building exterior profiles are stepped to give residences a terrace to appreciate the health benefits of sun and nature. Quality amenities normally associated with a villa lifestyle. The effect reduces building volume and creates psychological building pressure relief. The building becomes "gregarious" in scale and appearance. Green shade and the scent of flower herald the spirit of the old villas in Shanghai to the residences.

The red tile roofed buildings with surrounding green spaces increases the real land value of the site, neighbouring properties and surrounding areas.

Site area:	14.4 ha
Gross floor area:	136,000 m²
F.A.R:	0.94
Total amount of units:	635

上海达安圣巴花园No.3
Da'an St.Babara Valley No.3

中国：上海 Shanghai, China
业主：上海达安泰豪置业有限公司
Client: Shanghai Da'an Taihao Real Estate Co.,Ltd.
设计时间 Design: 2007
建成时间 Completion: 2009

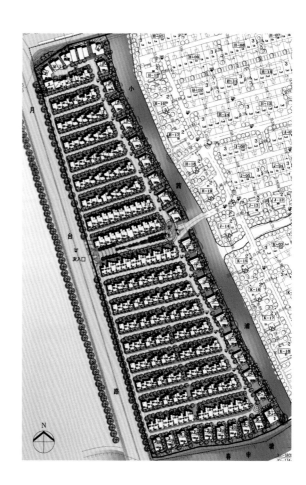

总平面图 Site Plan

本项目位于上海市松江区新桥镇，基地北临明中路，西临月台路，南临春申塘，基地西部有南北向河道小茜浦贯穿流过。该设计范围为它西侧的二期用地，是由明中路、月台路、春申塘、小茜浦围合构成的空间。地块狭长，长边为南北朝向。

本项目实际建设用地面积为4.3 ha.，地上总建筑面积30 000 m²，地下总建筑面积13 200 m²。建筑均为三层以下的联体住宅和少量独立住宅，沿入口处设少量配套低层公建设施。

设计的主力户型面积为90 m²以下的创新联体别墅。它不仅是安居工程，同时也可以成为小康社会很人性、很优美、很舒适、结合室内外空间的优秀居住建筑典型。
设计原则：
1. 所有屋顶以小坡屋顶为主，且不得相连。
2. 每户之间有后退，形成空间，种植高大树木与竹子在立面上形成独立的感觉。
3. 拉开与传统联排住宅立面横向线条作法的距离，强调竖向线条、丰富色彩、相邻立面均不相同，体现独立性和个性。
4. 每层后退，形成平台、露台。立面窗户放大，符合当地的居住需求，但不相同，上框或下框对齐。
5. 建筑与环境一起设计，花坛、花钵、木艺、铁艺处处与建筑结合，相得益彰。
6. 采用框架结构，抗震性能好，室内无承重墙。
7. 100%及以上的得房率。前庭后院，居住空间从内到外的延伸。中庭空间、内院空间，自我创意空间无限。

90花园别墅是KFS设计公司的中国国家专利产品。

用地面积：	4.3 ha.
总建筑面积：	30 000 m²
容积率：	0.70
住宅总户数：	268

The site is located in Xinqiao Town in Songjiang District in Shanghai. The site has a long and rectangular of shape due to proximity to the Xiaoxipu River that crosses the site from south to north.

The total site area is 4.3 ha. The gross floor area is 30,000 m². The overall housing program consists of 3-storey townhouses and a small number of detached houses. In addition to the housing program low level public facilities are situated near the entrances.

The new townhouses are 90 m² of architectural innovation. The tow house is more than an economic housing project. They are excellent designs for small houses and they offer a very comfortable living condition. A number of design principles including sloped roofs and not connected, and each unit has a yard suitable for vegetation. This increases the individuality of each unit. Emphasis on colorful vertical lines, instead of horizontal ones, makes the facades more personal and distinctive from each other. Using elements, such as vides, to make each floor within the same unit unique. A parallel design process is applied to both the units and the landscape. The usable floor ratio of more than 100% has greatly increased the value of the program.

Site area:	4.3 ha.
Gross floor area:	30,000 m²
F.A.R:	0.70
Total amount of units:	268

上海达安崇明御廷——90花园别墅
Da'an Royal Garden—90Villa, Shanghai

中国，上海 Shanghai, China
业主：上海达安锦迪房地产开发有限公司
Client: Shanghai Da'an Jindi Real Estate Co.,Ltd.
设计时间 Design: 2007
建成时间 Completion: 2010

总平面图　Site Plan

项目位于崇明县城桥新城，基地北临花鸟路，西临东引路，南临玉环路，东临鼓浪屿路。项目建设用地8.2 ha.，地上总建筑面积78 000 m²，地下总建筑面积17 800 m²，容积率0.95。规划为多层及低层住宅组成的居住小区，配置少量社区服务用房。

布局上将住宅划分为四种不同的产品类型：多层、联排、独幢和双拼。其中联排产品使用的是"90花园别墅"——KFS公司的专利产品。

综合考虑空间形态、交通组织和景观资源等多项因素，将几种住宅产品由西向东依次成组团排布。西侧临东引路为多层住宅，中间设置联排住宅，东侧临鼓浪屿路为独幢住宅，在鼓浪屿路上形成近低远高有层次的城市空间形态。各类型的住宅交通组织相对区分，主次有序。西侧多层住宅区的大面积集中绿化与东侧鼓浪屿路边的30 m绿化带相互渗透呼应，为整个小区构成均衡有机的绿化体系。

用地面积：	8.2 ha.
地上总建筑面积：	78 000 m²
容积率	0.95
住宅总户数：	783

The site is located in Chengqiao Town, Chongming District Shanghai. The project area is 8.2 ha. with a gross floor area of 78,000 m², The plans incorporate multi-storey apartment and low-rise residential units and a small amount of community service space.

Based on the planning, the residential program is divided into four different residential types, multi-storey, town house, detached house and semi-detached house. The town house is a KFS patented product for a 90 Villa.

Considering the factors of space, transportation systems and landscape resources, several houses are placed in block form from west to east. There are multi-storey apartments near Dongyin Road, and detached houses close to Gulangyu Road. The townhouses are positioned in the middle of the site. The sky line is outstanding clearly defined from east to west. The transportation system is organized according to an inherent logic. The entire green landscape streches between the multi-storey residential area in the west and the 30m green belt at Gulangyu Road.

Site area:	8.2 ha.
Gross floor area:	78,000 m²
F.A.R:	0.95
Total amount of units:	783

上海绿地南汇布鲁斯小镇——90花园别墅
Bruce Town—90Villa, Shanghai

中国，上海 Shanghai, China
业主：上海绿地集团
Client: Shanghai Greenland Group
设计时间 Design: 2007
建成时间 Completion: 2010

总平面图　Site Plan

项目地块位于上海市南汇区惠南镇，基地北临迎熏路，西临浦东运河，南临汇园路，东临南团公路。老港河从地块中间穿过，将地块分为南北两个部分。项目建设用地8.6 ha.，容积率1.3。规划为小高层及低层住宅组成的居住小区，并配置社区服务及商业用房。

设计师将本项目定位为惠南的一座都市田园小城，一个品位高档、环境幽雅、功能结构完善的生态社区，并努力实现以下几项目标：

1. 将面积跨度大的不同户型住宅有机组合，形成复合型和谐社区。

2. 努力开创花园式居住环境，合理配置开放性和私密性景观，令居民享有安详和谐的居住氛围。

综合考虑了空间形态，交通组织和景观资源等多项因素，将三种住宅产品由北向南依次成带状及片状排布。总体上形成南低北高的城市空间形态。各类型的住宅交通组织相对区分，主次有序。浦东运河与老港河河道景观以及小区内部的庭院绿化相互渗透呼应，为整个小区构成均衡有机的绿化体系。

其中联排产品使用的是"90花园别墅"——KFS公司的专利产品。

用地面积：	8.6 ha.
总建筑面积：	106 800 m²
容积率：	1.3
住宅总户数：	721

The site is located in Huinan Town in Nanhui District Shanghai and is surrounded by four major traffic arteries. The Laogang River divides the land into north and south parts. The total area of the site is 8.6 ha. with a floor area ratio of 1.3. The master plan consists of mid-rise apartments, low-rise residential units, service facilities and retail blocks.

The design approach provides for creation of an urban garden town with an elegant, prestigious and high social atmosphere. The combination of the apartment units of different size and modules is conducive to a multi-social mix. The creation of a garden-like environment with public and private landscapes results in a peaceful and relaxed atmosphere. The construction of the landmark introduces the site and its unique character to the city.

Following the logic of the geographical under layer, the transportation network and the landscape layer, the three modules are placed as bands with a north-south orientation. The mid-rise apartments are located at the north of the site, close to Yinxung Road. The majority of the 180 m² townhouses are situated along the Laogang River and function as an urban belt wrapping the smaller sized town houses. The skyline increases in height from south to north. The transportation system is organized according to an inherent logic. The entire green landscape streches between the water features and the inner landscapes.

Site area:	8.6 ha.
Gross floor area:	106,000 m²
F.A.R:	1.3
Total amount of units:	721

上海静安达安花园（部分一期除外）
Da'an Garden, Shanghai (Excludes part of phase 1)

中国，上海 Shanghai, China
业主：上海市达安房产
Client: Shanghai Da'an Real Estate Inc.
设计时间 Design: 2001
建成时间 Completion: 2005

总平面图　Site Plan

达安花园位于上海市中心静安区长寿路武宁路口，设计依基地及方位条件，将32 ha.的住宅区清晰地分为三个组团。各个组团围绕各自的组团绿地（5 000～8 000 m²），三组团又环绕 约9 000 m²的中心花园，取得户户朝南，家家面绿的设计效果。空间归属感强，绿化生态环境优越，极佳地补充了现代城市生活所缺少的居住氛围。围合式空间的创造使住户在大型住区中更易于识别属于自己的生活空间，在视觉和心理上易于产生认同感和安全感。住户单元出入均安排在组团花园周围，使花园更贴近住户的日常生活。达安花园在长寿路和武宁路各设有一主要出入口，余姚路上设辅助入口。连接两个主入口的车行道及带形密林花园，是小区主要的东西轴线，具有重要的指向性和识别作用。

为充分体现这一长约200 m，宽约25 m的带形空间的生态功能，将原布置在林带中央的会所转移，成为一条纯粹以绿树为主的生态林带，突出达安花园的生态主题，同时绿色的轴线也为居民带来更多的享受。而原小区会所将替代中心花园中购物亭等小品建筑的位置，并采用半地下的方式缩小体量感，以结合中心花园景观之建筑形象成为实用的小品建筑，并以其活泼优美、细节丰富的玻璃体建筑成为中心花园的组成部分，既发扬了原有的绿色主题，又为会所的使用者提供广阔、丰富的环境因素，使总体的设计价值又得以提升。

三个组团绿地下方均设有地下车库。为改善车库内日常通风、采光条件，同时为平缓的绿地制造一些变化，在车库的侧向或中央均留有大型采光面或采光井，结合环境设计处理成山坡或瀑井，更丰富了组团内的景观，体现了本设计以功能为原则，发掘一切可能的景观条件，创造独特、优美生活空间的宗旨。

用地面积：　　　　　　　　　9.0 ha.
总建筑面积：　　　　　　　　310 000 m²
容积率：　　　　　　　　　　3.4
住宅总户数：　　　　　　　　2 865户

Two major traffic arteries, Wuning and Changshou Roads, and a minor street enclose the triangular property located at the west end of the Jing'an District in downtown Shanghai. The residential development contains 3,000 families and a junior high school.

Three half enclosed residential groups are designed, each with 5,000 to 8,000 m² gardens and a 9,000 m² central park. This softens the negative impact of a dense urban lifestyle through the reduction of unwanted noise and noxious odours from traffic, and addresses a primary design consideration needed for small scale living spaces. Residences have entrances from the central garden path and each apartment is located near a park that gives a welcomed sense of security, neighborhood identity and community spirit.

Three main entrances are located on the streets bounding the site. The west and the east entrances connect a 25 m wide and 200 m long treed park, for a green axis on the property. This provides a desirable leisure space in a green environment. The third entrance is near the clubhouse, equipped with a gym and an indoor swimming pool, and a recessed forming a central park setting and a good view of the vegetation. This allows more open space for recreation.

Underground parking garages are located beneath the three gardens. The distinctive feature of the garden design is the integration of the garage "light wells" that are either surrounded by waterfall features or, become playground areas. The high desirable quiet, green living spaces in a dense metropolis create a successful commercial real estate market development.

Site area:　　　　　　　　9.0 ha.
Gross floor area:　　　　 310,000 m²
F.A.R:　　　　　　　　　　3.4
Total amount of units:　　2,865

上海静安达安锦园
Da'an Jin Garden, Shanghai

中国，上海 Shanghai, China
业主：上海达安房产
Client: Shanghai Da'an Real Estate Inc.
设计时间 Design: 2001
建成时间 Completion: 2008

总平面图 Site Plan

达安锦园位于上海中心区常德路、西康路、安远路与海防路形成的街坊内，基地位置交通便捷，市政设施完善。设计构思主要从基地自身特点出发，找寻来自内部和外部，自然和人文的诸多可利用因素，综合得出一个适合于该地块特征的新时代住宅小区。提倡与自然之间相互谐调、相互融合的理念，突出"以人为本、为人服务"的设计思想，创造出一个既能利用城市各种便利，同时又能享用自然恩惠的健康、优美的理想居住环境。

在总体布局上，最大可能利用城市绿地，同时又创造出优美的小区内部环境，使大部分住宅有最佳朝向和最美景观。南块以及北块均为大围合的内向型空间，住户能同时享用精彩的内部环境和优美的外部环境。

在环境设计方面沿用了上海特有的海派文化和先进的环境设计理念，创造出美观、经济、实用的特有景观环境。

用地面积： 4.3 ha.
总建筑面积： 160 000 m²
容积率： 3.7

Da'an Jin Garden is a mid to high range residential development and is located near a triangular city park and north of the Jing'an District of Shanghai. The land has the advantage of a downtown location and existing established amenities. The site character of natural conditions and cultural features, control the design concept and create a modern, healthy and beautiful living environment that is in harmony with nature.

The city green space strategically arranges the units to provide them with the best views and sun orientation. The north and south blocks of the complex are designed to have an enclosed courtyard that gives each family a garden and an urban view next to generous amounts of sunlight.

The landscape design uses both traditional Shanghai and modern gardening techniques. The view of the city park is extensively utilized in the development.

Site area: 4.3 ha.
Gross floor area: 160,000 m²
F.A.R: 3.7

总平面图 Site Plan

上海普陀区愉景华庭位于上海西区万里居住区内。依据市区中少见的40 m宽中央绿地规划的特色，塑造宜人的居住环境是设计所追求的目标。

以规划绿轴为小区中心主轴，将各个住宅组团的围合空间依次展开。在保证中央绿轴的前提下，使这个40 m宽的绿轴贯穿小区中央，由新村路主入口以一个"瀑布栈桥"做序，奠定了主轴之首，而后蜿蜒向北，与小区北人行入口和会所结合。在这里，将有一座桥梁跨越横港河，与北部的大华居住区相连。绿轴的游动，加之各个组团的微曲体量围合，产生了无与伦比的动态美感，造就出变化丰富的总体环境与组团空间。

房型分布与总体关系息息相关，井然有序，其布局表达以下几个设计要点：
1. 沿绿轴布置较为高档的大房型；
2. 沿东西干道侧以小高层板式布局与内部高层形成对比，从而更加突出中央绿轴；
3. 沿富水河路布置较高层次板式楼，以南低北高格局统领大势；
4. 在西北角以独创塔式公寓楼成为小区至高点，与低层会所、商业一齐汇合成另一种对外界面。建筑形象设计以清新亮丽、通透轻盈为特点。住宅入口处，形成南北进厅布局，使底层门厅更为开敞，更有置身于绿野环抱之中的感觉。

用地面积：	9.1 ha.
总建筑面积：	227 000 m²
容积率：	2.1
住宅总户数：	1 630

The Yujing Garden in Putuo District, Shanghai has a 40 m green belt as a primary design that creates a comfortable living environment.

Clusters of housing are located next to the central located 40 m green way. A bridge over a waterway functions as the introduction to a landscape corridor, which is linked. by a series of pedestrian entrances, to a clubhouse at Fushui. Wanquan Road provides a gentle transition from a natural environment toward a man made built environment.

The masterplan has several design features including the most luxurious and large housing types located beside the green belt. The mid-rise buildings are along the east to west main road. This contrasts the high-rise buildings. These buildings emphasize and compliment the central green belt. The high rise buildings with their slab like character located along Fushuihe Road provide a magnificent 360 degrees view of the sky-line.

The " tower " high-rise buildings form the highest point at the northwest corner of the site. At the lower level they merge with the low-scale club and commercial facilities creating a new and alternative architectural image. This image is reflected by its outstanding facade design. The housing entrances open up the ground foyer to the surrounding green belt.

Site area:	9.1 ha.
Gross floor area:	227,000 m²
F.A.R:	2.1
Total amount of units:	1,630

上海黄浦明日星城（部分一期除外）
Tomorrow Star City, Shanghai (Excludes part of phase 1)

中国，上海 Shanghai, China
业主：上海东方金马房地产发展有限公司
Client: Shanghai Orient Golden Horse Co.,Ltd.
设计时间 Design: 2001
建成时间 Completion: 2005

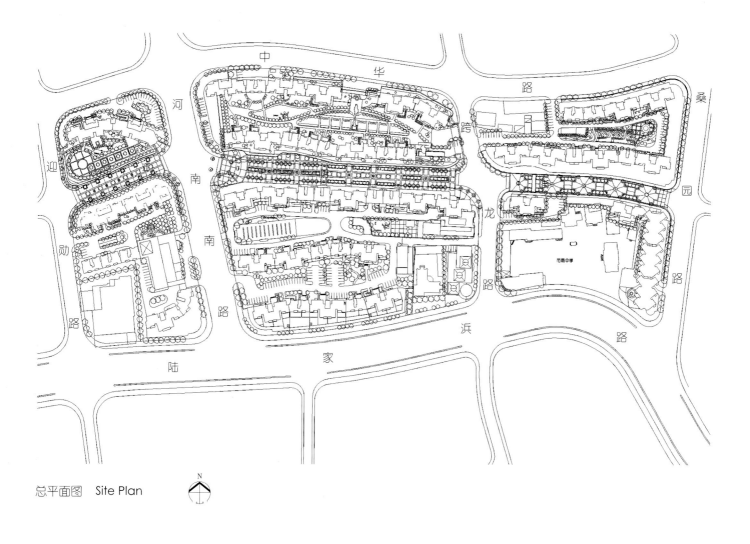

总平面图 Site Plan

明日星城位于上海市黄浦区老城区中心，整个基地被城市道路分隔成六个街区，是以住宅为主，集商业、影视娱乐、教育文化等多种功能为一体的城市型住宅区。设计力求在空间布局、住宅风格、室内外环境上创造具有一定品质的高尚生活环境。总体布局在极为紧凑的用地内尽可能多地布置了休闲绿地，以营造闹中取静的居住氛围。高层住宅基本上均沿着道路布置，沿街住宅底部设商店并利用商业裙房将小区内部新绿地围合成封闭式的内院，形成一个与喧闹的街道隔离开来的安静的居住庭院。把多幢住宅视为整体立面来设计，给城市道路留下一个连续舒展的立面形象。

住宅房型均为经济型的二房或三房，以控制每套房子的总价。沿江阴街步行街则布置有室外茶座，增加了步行街的人文生活气氛。建筑主要以18～35层的高层住宅为主，不同层数相互组合，使建筑体量呈现出高低错落、变化丰富的外观效果，并采用曲线形的拼接方式，使建筑形象含蓄而又新颖，在城市天际线上形成了流动、连贯、寓变化于群体形象中的清新格调，具有很强的个性特点和可识别性。

The urban renovation development property is north of Lujia-bang Road, west of Songyuan Road in the Huangpu District of Shanghai. The master plan contains residence, theatres, and education facilities. The design emphasizes the high living standards for a new century lifestyle and creates a contemporary and "upscale" quality environment. Semi-enclosed spaces with gardens create a secluded "garden type" space in the city. Low buildings are arranged adjacent to the major traffic artery to reduce the visual pressure to the residents.

Open seating coffee and tea places are randomly placed along the Jiangyin pedestrian street for a comfortable living and relaxation character. 18 to 35 storeys residential buildings are linked in a slightly curved manner to form a vivid fresh harmonious image. The high profile development, distinctive in appearance, offers a major architectural design statement to the neighbourhood.

Site area:	11.1 ha.
Gross floor area:	450,000 m²
F.A.R:	4.0

规划用地面积:	11.1 ha.
总建筑面积:	450 000 m²
容积率:	4.0

上海浦东东晶国际公寓
Dongjing International Residential, Shanghai

中国，上海 Shanghai, China
业主：上海东道置业有限公司
Client: Shanghai Dongdao Real Estate Inc.
设计时间 Design: 2002
建成时间 Completion: 2007

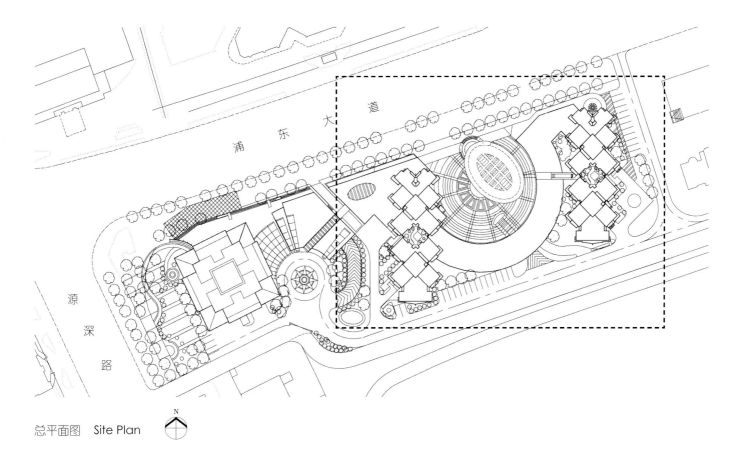

总平面图　Site Plan

该基地位于上海浦东大道源深路，包含公寓、商业等功能。公寓楼以45度朝向的鱼骨架平面形式，巧妙解决日照等诸多矛盾，并避免使浦东大道临街面出现冗长、闭塞的建筑体量，同时保证了户户朝南的效果。设计房型为60～90 m²的小户型，而其中公寓中部的每五层高的共享中庭尤受客户欢迎。南部有"上海滩花园洋房"的绿荫深深的景观。曲面的商业建筑集中布置在两栋公寓楼之间，沿街布置以中小型复式商铺为主的连续商业，有效围合并形成具有园林景观特色的外向型城市商业广场。

本方案意在从建筑文化和环境人文角度着手，用崭新的时代建筑语言，建设一个环境优美、设施完备、风格独特的国际化居住园区。多种功能既合又分的布局，实现多赢格局。极富韵律、简约的总体布局以及戏剧性的构图，使得空间层次更加丰富。现代的建筑外观及商业建筑明亮通透的立面造型，在浦东大道上让出更多的空间，留下更多的阳光，使其对城市主干道的压力大为减少并提升内部空间品质，彰显小区的国际化概念。

The development is situated on Pudong Avenue, east of Yuanshen Road in Shanghai. The design development program of commercial and residential was refined and honed to encourage the mutual benefit of the different functions. A new architectural style introduces a design image vocabulary of an international multi-functional complex within a pleasure environment. The retail space, composed of small and medium sized shops along the street, is located between the residential buildings that enclose an open urban pedestrian plaza with a complementing landscape feature.

Due to a lack of land area and a high density the design maximizes the amount of open space where people can enjoy nature and sunshine. The design of a two-storey atrium, open retail plaza, elevated platform, a five-storey sky garden and a roof garden greatly enhances the interior environment and relieves the high pressure fast pace of the Pudong Avenue arterial road.

用地面积：	1.7 ha.
总建筑面积：	66 000 m²
容积率：	3.8
公寓总建筑面积：	33 000 m²
住宅总户数：	452

Site area:	1.7 ha.
Gross floor area:	66,000 m²
F.A.R:	3.8
G.F.A of apartment:	33,000 m²
Total amount of units:	452

上海长宁春天花园
Spring Garden, Shanghai

中国，上海 Shanghai, China
业主：上海东方金马房地产发展有限公司
Client: Shanghai Orient Golden Horse Co., Ltd.
设计时间 Design: 2001
建成时间 Completion: 2005

总平面图 Site Plan

"春天花园"位于上海市长宁区长宁路,临近虹桥开发区。设计以错位经营的思维模式为指导,创造了与周边地块、土地性质相同而产品定位差异化的新型住宅小区。整个160 000 m²的住宅区分成了一半5层的叠加别墅,一半18~30层的小高层及高层两大类形式。在2.5的容积率地块上,开发出了本地区绝无仅有的4~5层叠加别墅房型,成为稀缺品种而凸显了价值。不同类型住宅带来的建筑体型变化,丰富了沿长宁路主要干道沿线的城市天际线,赢得了城市规划部门和消费者的肯定。

叠加别墅的1~2层是复式,拥有地面南向的庭园,3~5层又是一个三层复式,顶层退台设计了大型的屋顶花园,并有一个全玻璃顶的"星光室",以使久居城市之人在室内也能不时感受天空和自然的魅力。半地下室一部分是楼上住户的停车库,还剩下一些空间给1~2层的住户作储藏室之用。这样造就了类似于别墅的生活方式和大量可用的空间。

用地面积:	6.5 ha.
总建筑面积:	160 000 m²
容积率:	2.5
住宅总户数:	1 177

Spring Garden is situated in the Changning District of Shanghai, south of the famous Hongqiao Development Zone and the Tianshan commercial centre. With Suzhou Creek on the north, Zhongshan highway and Zhongshan Park on the east and a new high-rise residential development on the west, the property has excellent residential development prospects. A market analysis of the residential real estate trend concluded with a design of both high-rise apartment towers and low-rise double-townhouses that offer different residential life style choices. Double-townhouses, two vertically arranged double-height units, recall a traditional Shanghai 'lilong' house with small courtyards and compact two-storey living. Gardens and spaces on a more intimate scale are designed wherever possible, bringing nature back into the city life.

The residences to the south, and high-rise towers on the north bring maximum sunshine and views of a central garden. An entrance boulevard and a clubhouse along the main garden combine contemporary western and traditional Chinese garden features. The trees, grasslands, streams and fountain features present many different garden experiences. The street side building ground floors are elevated to bring the garden into view from the city making the Changning District a new unique urban design experience.

Site area:	6.5 ha.
Gross floor area:	160,000 m²
F.A.R:	2.5
Total amount of units:	1,177

上海虹口明佳花园
Mingjia Garden, Shanghai

中国，上海 / Shanghai, China
业主：上海明佳房产开发有限公司
Client: Shanghai Mingjia Real Estate Co., Ltd.
设计时间 Design: 2001
建成时间 Completion: 2005

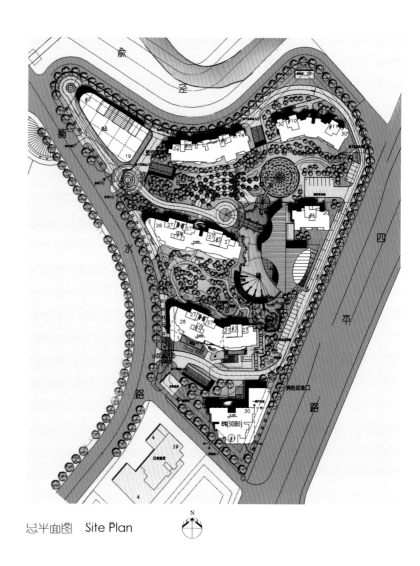

总平面图 Site Plan

明佳花园位于上海市虹口区中心腹地，紧临四川北路商业街及新建城市绿地，交通便捷，市政设施完善，一期地块沿大绿地展开，拥有很好的景观条件。

虹口区人文气息浓厚，具有独特的文化品位与传统，所以总体设计意欲在创造良好空间和生态环境的同时，传承和发扬地区文化之精华，设计具有文化品位的高尚居住生活区。小区在总体布局上，最大可能地将更多的住宅单元面向四川北路大绿地，以沿衡水路展开的舒展、外向型空间布局，使大部分住宅既有最佳朝向，又有最美景观，并结合重点的立面处理，使之成为建筑群的视觉焦点。

小区环境最大限度借用四川北路绿地景观，把住宅区景观设计融入大绿地的组景之中，使两者互为补充与延伸。建筑的立面上，打破高层板式建筑固有的巨大体量，立面处理分散了体量，减轻建筑顶部的重量感，使建筑形象轻透、体量变化丰富，并以一系列垂直的线条和顶部的各种处理形成一系列组合，避免了大体块的视觉压抑感，在开敞绿地旁制造一种轻松、优美的居住建筑形象。同时建筑细部尽量运用玻璃、轻钢装饰以弱化大型建筑体量带来视觉上的压抑。

Mingjia Garden is a residential development in the Hongkou District of Shanghai. Located at the centre of the district, adjacent to Sichuan Road commercial centre and New City Park, the property has a great park view and access to all the urban amenities.

Hongkou District has a rich cultural heritage. The development design concept creates a comfortable living space and restores the unique cultural heritage with a modern architectural style. In the park the residential towers face the best sun orientation and ensure that most of the residences have a park view. The park-like natural setting is extended into the development with high verandas and porches that offer exterior sitting areas in a garden setting. A clubhouse beside the Siping Road, a busy traffic artery is covered with a landscaped roof sloped toward the garden that acoustically dampens the ubiquitous traffic street noise.

Bright and spacious living and dining rooms are in the centre of the residences offering the best views. The building are designed as slender towers to reduce the visual impact, and support the delightful openness and park-like setting. Extensive use of glass and metal with architectural glazing features of low sill windows, bay windows and corner windows that bring the impression of grace and elegance to the buildings.

用地面积：	2.5 ha.
总建筑面积：	100 000 m²
容积率：	4.0
总户数（含居住办公一体建筑）：	667

Site area:	2.5 ha.
Gross floor area:	100,000 m²
F.A.R:	4.0
Total amount of units(including SOHO):	667

上海徐汇电影华苑
Cinema Garden, Shanghai

中国，上海 Shanghai, China
业主：上海东方金马房地产发展有限公司
Client: Shanghai Orient Golden Horse Co.,Ltd.
设计时间 Design: 2001
建成时间 Completion: 2007

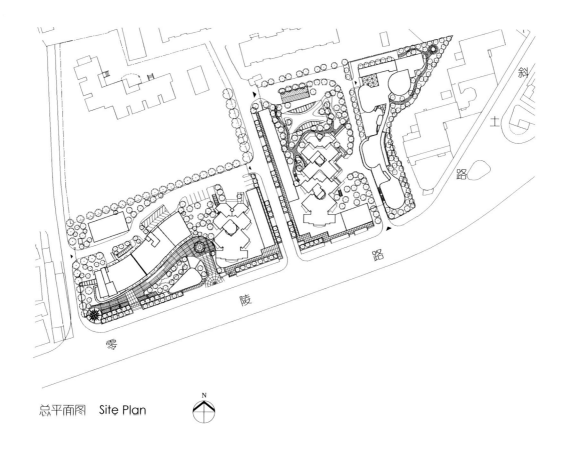

总平面图　Site Plan

电影华苑基地为上海电影制片厂旧址，紧邻上海徐家汇八万人体育场和上海体育馆。设计构思尽可能地与周边环境相协调，去弊取利，形成唯一性的建筑布局形式。设计房型以60～80 m²的小户型和出租性公寓为主，并设有一个外向型的小型商业步行街。公寓会所设在地下室，包括一个室内恒温游泳池。这些配套设施将极大地满足住户的日常生活的需求。

高低错落的建筑布局，既是基地日照条件的限制，也给城市带来丰富的天际效果。利用地下停车顶板低于路面0.6 m堆土造景，组织跌水涌泉，结合绿篱、草地、雕塑等共同塑造了优美的户外生活空间。每套住宅均力求创造明亮整洁、舒适宜人的居住空间。

建筑的立面采用大量玻璃栏板阳台及大面积低窗台凸窗，立面由下至上从实至虚，沿街商业裙房为一系列精致挑篷，充分体现了简洁优雅、透亮的建筑风格。

用地面积：	1.9 ha.
总建筑面积：	66 000 m²
容积率：	3.5

The development is north of the Lingling Road in the Xujiahui District of Shanghai. Being aware of the existing streets, traffic patterns, and pedestrian flow, the design blends the new towers into the surrounding neighbourhood with design concepts for daylight hour operation and strong pedestrian pattern routes. High rise buildings step in height with a vivid cluster of towers offering all residents access to natural light, landscaped gardens with ponds, waterfalls, green barriers, lawns and sculpture.

Site area:	1.9 ha.
Gross floor area:	66,000 m²
F.A.R:	3.5

建成时间 Completion: 2005

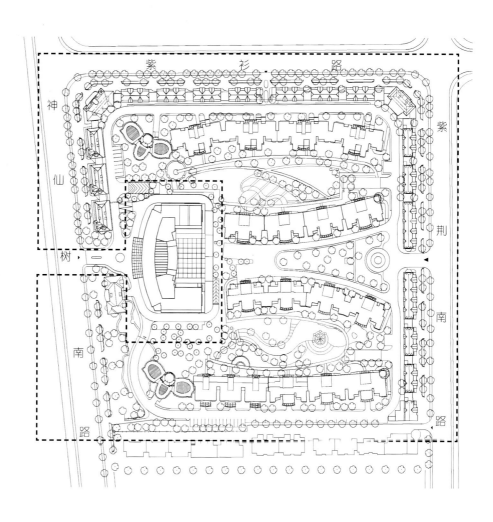

总平面图　Site Plan

"上海花园"是建于成都市高新区神仙树的高档住宅小区。基地东临紫荆南路，西靠神仙树南路，南临近机场路西延线，北为紫杉路。整体项目是一个规模适宜，以住宅为主、商业配套为辅的高档住宅小区。

基地内五星级酒店大大提高了本小区住宅的价值。大量的小高层建筑配以沿街上海小洋楼式的商业建筑，是开发商上海情结在成都这片热土上的集中反映，是优雅和格调的诠释。通过总体布局及建筑立面造型的层次演进，达到相辅相成、相互辉映，若即若离，互不干扰的美好境界。

用地面积：	5.8 ha.
总建筑面积：	145 000 m²
容积率：	2.5

Shanghai Garden is an upscale residential development bounded by South Zijin Road to the east, South Shenxianshu Road to the west, Jichang Road to the south and Zishan Road to the north in Shenxian Shu District in south Chengdu. It is a suitably mixed use scaled residential commercial development.

The inclusion of a five-star hotel on the property substantially increases the real estate value of the four "mid to high" level houses. Residence owners participate in administration profit sharing. The Shanghai developer brings the old Shanghai style to the development with the return of elegance and beauty to the design. Three different styles in the master plan are integrated with one another and form complete body that results in a harmonious environment.

Site area:	5.8 ha.
Cross floor area:	145,000 m²
F.A.R:	2.5

上海绿地成都维多利亚花园
Victoria Garden, Chengdu

中国，成都 Chengdu, China
业主：上海绿地集团
Client: Shanghai Greenland Group
设计时间 Design: 2004
建成时间 Completion: 2007

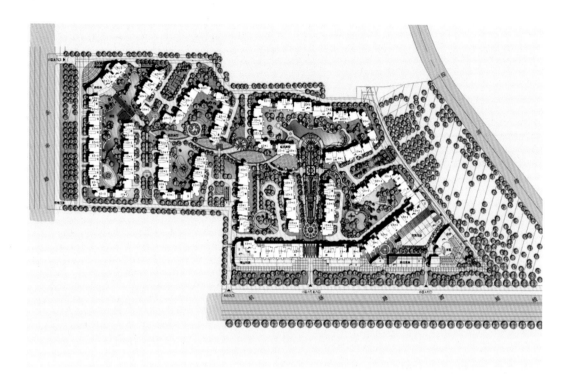

总平面图　Site Plan

本项目地处四川成都高新区，如何运用适合当地气候特征和文化传统的规划设计手法，创造具有地方特色与自然环境相协调的现代高尚生活社区，为居民提供健康舒适的居住生活环境成为本项目的重要切入点。基于"道路外环，人车分流"的设计构思，住宅建筑利用半围合式布置形成一个个大小不一的半开放式院落空间。会所、农贸市场设于基地东南角处，南侧沿机场路设置了商业用房，创造出生活领域内置型的理想型居住环境。同时各种庭院与建筑巧妙地组合、配置，形成了一种随季节不断变化的居住场所。

根据建筑总体规划布局的特点，运用自然造景手法，结合欧洲园林特色，以山林主体为构架，形成一个浑然一体的山水住宅园林。通过现代的建筑手法和经典的欧式建筑元素，体现建筑深厚的文化底蕴及尊贵的社会形象。

居住区规划车行入口位于西侧与南侧，在庭园外侧环通。同时，南侧有一个人行景观入口。尺度亲切的水池，精致的铺装，引领着人们进入一个个步移景异的院落之中。住宅围合的花园、底层架空的空间与住宅自身的空中花园，使风景与建筑融为一体，室内与室外的界定被模糊。自然和风景的连续性使自然景观轴线与规划景观轴线自然融合，创造了一派休闲的田园风情和气氛。

The site is in the New Development District of Chengdu, east of the Yongfeng Road and north of the Jichang Road, with a background of intense urban construction, diverse cultural heritage, and important modern thinking. The critical development issues are a healthy and comfortable environment, a vernacular in harmony with nature and the local climate within the area heritage context. The separation of pedestrian and vehicle transportation concept, supports the building layout from the east to west, forming a number of half open courtyards. The clubhouse and the vegetable market are on the southeast with commercial spaces designed along the Jichang Road on the south. The design creates a commercial and residential separated community.

The master plan contains natural style landscape that elevates the community life to a higher level and introduces geometrical forms. Framed by the terrain, the green spaces mutually intersect and form the integrated community garden. The architecture exhibits a noble image and introduces western classical elements.

Two vehicle entrances are on the south and west, and a pedestrian entrance leads people into different courtyards with pools and pavement on the south. The soft natural gardens and raised ground floor space cushion the hardness of the interior and exterior space. The continuity of natural landscape with planned landscape axis generates a quiet, idyllic atmosphere.

总用地面积：	12.1 ha.
地上总建筑面积：	290 000 m²
容积率：	2.5
住宅总户数：	1 718

Site area:	12.1 ha.
Gross floor area:	290,000 m²
F.A.R:	2.5
Total amount of units:	1,718

中国，杭州 Hangzhou, China
业主：杭州瑞城房地产开发有限公司
Client: Hangzhou Ruicheng Real Estate Co.,Ltd.
设计时间 Design: 2003
建成时间 Completion: 2009

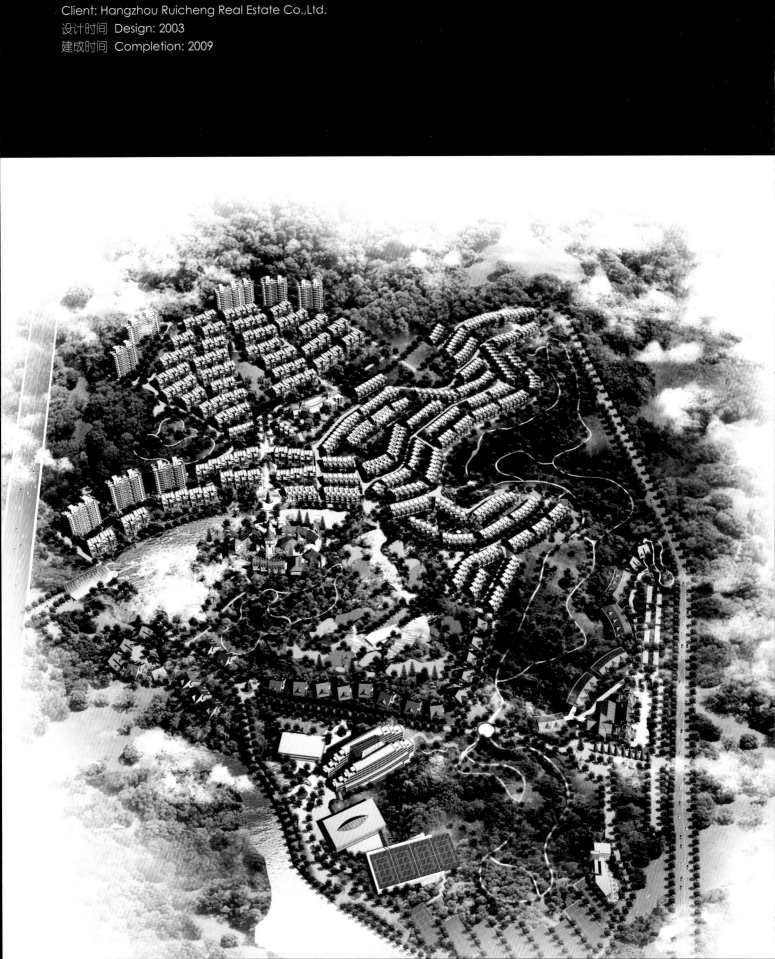

总平面图　Site Plan

基地位于杭州市余杭区仓前镇宋家山村，地块周边有西湖风景区、径山风景区等众多旅游资源。基地地形以山坡丘陵为主，北临凤兴路、南靠宋家山、东接良上公路、西接东西大道。小区主出入口设置在良上公路和东西大道上。本案用地面积20.3 ha.，总建筑面积约为270 000 m²。

设计上依托项目独有的地貌，一直贯彻"有山有水"的居住小区及体育酒店休闲"圣地"的主题思想。

居住地块位于北部，依托自然山水创造生态型居住小区，因地形或房型的不同而区分成若干个组团。旅游地块位于南部，引入以网球运动中心为主题，以会议中心度假村为载体的"健康运动休闲旅游区"概念，突出了休闲旅游项目的主题内容，提高了人气。居住区封闭式管理及旅游区的开放模式恰恰适应这种功能互补的需要，从而有效减少重复设置相关设施。环境设计上尽可能保持其自然的特色。

岛的东、北、西三面环水，环顾四周，视线极为开阔。建于此地山顶的欧式城堡会所，为居住区以及体育设施提供饮食、娱乐服务，也是整个小区主要景观的要点。除了在"梦庭"入口处可以观赏到山顶城堡外，在小区居住地块内及旅游地块的北部，它都是视线的焦点所在，也形成一个依山傍水的建筑人文景观中心。

用地面积：	20.3 ha.
总建筑面积：	270 000 m²
容积率：	1.2
住宅总户数：	1 842

The development site is located in Hangzhou city. The terrain of the site is slight hills and surrounded by different urban elements. The two main entrances of the site are located at Liangshang Highway and Dongxi Avenue.

Governed by a unique geographical condition, the design concept features "water and mountain" and "sport resort" as primary design parameters. The residential area on the north of the site is in the ecological zone adjacent to the natural environment. The residential precincts are divided into several groups based on the landform variations and residences. The tourist area on the south, introduces the conference and tennis courts as the "healthy sport" resort concept that promotes the professional prestige of the development. The closed residential quarter and the open tourist area conveniently avoid repeated intervention of service facilities. The natural scenery is preserved and the local mountain water is used.

Site area:	20.3 ha.
Gross floor area:	270,000 m²
F.A.R:	1.2
Total amount of units:	1,842

中国，上海 Shanghai, China
业主：上海汇成房产经营公司
Client: Shanghai Huicheng Co.,Ltd.
设计时间 Design: 2002
建成时间 Completion: 2007

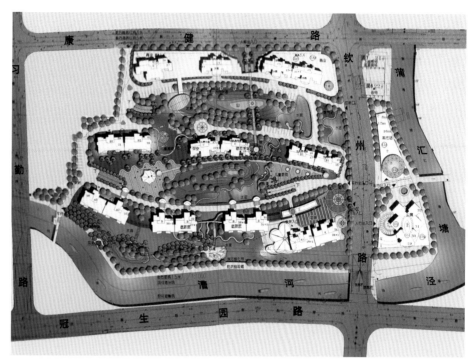

总平面图 Site Plan

漕河景苑项目位于上海徐汇区漕河泾街道中心，北邻康健路，东靠蒲汇塘，西临习勤路，南依漕河泾港。小区以住宅开发为主，辅以商业、会所等居住配套设施。小区楼盘规模适中，地理环境得天独厚，它将以一个令人耳目一新、心旷神怡的都市水景住宅展示在世人面前。基地南侧、东侧为宽阔水域，基地西距康健公园约500 m，具有一定的景观优势。基地所处区域环境较为幽静，整体环境优良，周边交通发达，出行方便。

设计最大限度利用基地所处漕河泾河道水景，将自然水面与人工水面巧妙结合，形成既得天独厚、又浑然天成的全水景住宅。设计还将建筑的景观与水景结合，使得小区的建筑景观也能成为城市的一道亮丽风景线，达到城市规划需求与住宅开发效益双向共赢。基地东部是相对独立的青年公寓，视觉景观极其开阔，其挺拔、隽秀的立面造型，成为半岛形基地上的地标式建筑。

沿河岸活泼地布置了二排精致、典雅的低多层住宅，减轻对水体空间上的压迫感，更使小区与城市河道景观融合共生，相映成趣。西侧的高层住宅与晶莹的会所在星空下的水面上产生美妙倒影，映衬出美丽的沿江风景。而其南侧的河道，则有绿岛、沙滩与水生植物组成了绿色的沿江生态景观。小区的中心是以自然手法营造出山水相济的中心绿地，使该组团住宅享受南北双向景观。小区的北部是有便利商业设施的南向大纵深观景房。设计以基地内在地域文化为根基，引发深刻含义的审美情趣，提升小区的审美层次。

The project is the design of a residential area in Downtown Shanghai. The site has outstanding geographic terrain features including a reservoir basin of fresh water. The landscape design is perfectly harmonized with the river at the south and east of the site. The surrounding environment is quiet with different modes of transportation available.

The exceptional beautiful waterscape integrates the existing natural waterscape with the implemented man-made waterscape. The design plans for residential buildings with full waterscape views. The landscape surrounding the residential buildings integrates land and water. This makes the area a beautiful scenery and valuable addition to the city. The design takes the urban planning requirements as beneficial for the building development. The service apartment typologies are located in the east of the site. Their open views and location in a peninsula like landscape make them the landmark of this design.

Two elegantly draped lines of low-rise buildings integrate the riverside landscape with the landscape of the project. This creates a pleasant place for the local residents. The high-rise buildings and crystal clubhouse find their star-like reflection in the water of the beautiful river scenery. The ecological waterscape in the south consists of green islands, soft sand beaches and rivers rich with vegetation. The area in between the hills and the water integrates and draws on them as desirable qualities for the residents. There are, in the north, large module resort apartment buildings with additional commercial functions. The architectural design principals give a boost to the appreciation of aesthetics because they are based on local geographical conditions.

用地面积：	4.5 ha.
总建筑面积：	113 000 m²
容积率：	2.5

Site area:	4.5 ha.
Cross floor area:	113,000 m²
F.A.R:	2.5

上海宝山旭辉依云湾
LA BAIE D`EVIAN, Shanghai

中国，上海 Shanghai, China
业主：旭辉集团
Client: CIFI Group
设计时间 Design: 2002
建成时间 Completion: 2008

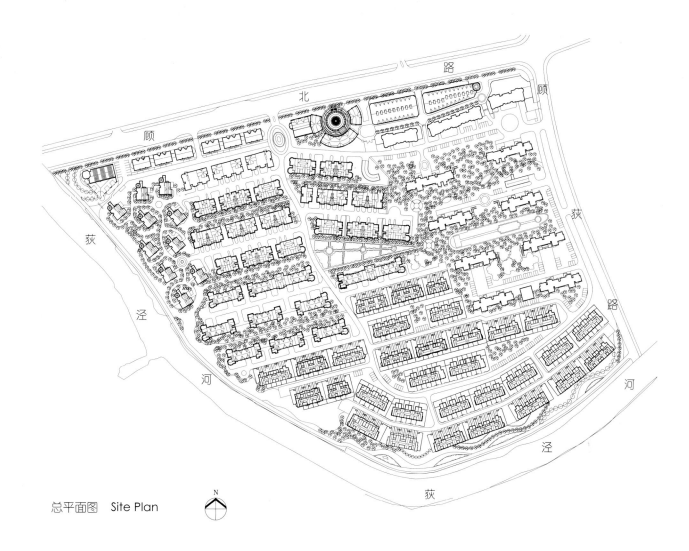

总平面图　Site Plan

基地位于上海市宝山区顾村地区，用地面积约19 ha.，总建筑面积约为220 000 m²，基地北临顾北路，东临顾荻路。南面和西面的用地分界线为狄泾河。同时，基地西侧的沪太路隔着荻泾河与本案两两相望。

设计借助鲜明、古典、华丽高贵之风雅底蕴与设计理念，以概括洗练的规划布局，独到精致的建筑设计，赏心悦目的景观意象，使此项目从不起眼的地段与不经意的环境中脱颖而出，成为宝山地区近市中心地块的一颗耀眼明珠。

住宅产品主要为板式高层和联排别墅。基地不同方位的地块特点决定了各种产品的布置。沪太路为城市道路，车流多、噪声大，沪太路与基地关系是越向北则两者距离越远。因此在基地西北区域的沿河流域，设置独栋别墅。

顾北路为小区主要入口，而且是主要车行道路，因此将三层的商业配套建筑集中设置在顾北路沿街面。

板式高层区设置在小区东面，并在顾荻路设有独立入口，使高层区和别墅区互不干扰。整体空间由西向东升高，保证荻泾河这一主要景观资源能被充分利用。

用地面积：	18.9 ha.
总建筑面积：	220 000 m²
容积率：	1.2

The site is in the Gucun Area in the Baoshan District of Shanghai with a total site area of 18.9 ha. with 220,000 m² of G.F.A. The site is surrounded by major arterial roads. The Dijin River crosses the development property from south to west.

The concept adds brilliance, intensity and classical elements to the design. Refined architectural designs are clearly planned, and situated in an original and pleasing landscape design. This project has the potential of becoming a shining pearl in the centre of Baoshan.

The majority of the residential program is designed as apartment blocks and townhouses, that are spatially set apart. The intensity of the urban sounds declines toward the northeast corner. The detached houses are situated, close to the river, in the north west area.

The main entrance of the compound is located at Gubei Road and is part of the three-level commercial strip that faces the street.

The high-rise apartment towers are situated on the east side of the site. All the towers have their own entrance enabling an easy separation of user flows. In order to have the optimum use of the views, the height of the skyline increases from west to east.

Site area:	18.9 ha.
Gross floor area:	220,000 m²
F.A.R:	1.2

上海徐汇百汇苑二期
Baihui Garden Phase 2, Shanghai

中国，上海 Shanghai, China
业主：上海百汇房地产开发有限公司
Client: Shanghai Bai Hui Real Estate Development Co.,Ltd.
设计时间 Design: 2007
建成时间 Completion: 2013

总平面图 Site Plan

此项目为百汇花园二期的住宅部分，在进行户型优化的同时进行立面改造设计。

首先，由于本案地处上海市徐汇区南部，东临黄浦江，北靠龙华港，并与2010年世博会场址隔江相对，得天独厚的景观优势是本案的最大潜力所在，因此，景观资源最大化利用就成为此次方案调整的指导方针。

通常人们把能看到江面的住宅都称作江景住宅，尽管有时候必须伫立窗前，甚至探出窗外。但在这个项目中，设计师们提出舒适观景的新概念，例如，坐在客厅的沙发上或者躺在主卧室的床上即能尽揽江面的壮阔美景，这样才能体现出豪宅的优势和设计的细腻。要达到这一舒适观景的效果，就必须在房型设计上充分考虑户型和景观资源的相对关系，公共空间和私密空间的位置以及开窗的方式和阳台的位置，观景面的幅度等很多细节。

根据此类原则，设计师们对项目中各栋住宅的标准层面积进行了重新分配，改变了原来较为平均的状况，使各栋住宅的面积能与其自身景观资源的状态相匹配。

用地面积： 21.7 ha.
二期总建筑面积： 165 000 m²

Baihui Garden, is located at south of Xuhui District, in Shanghai, and faces EXPO 2010 pavilions divided by Huangpu River. The foundation of the design is how to fully utilize the huge potential of landscape.

The apartments has views to the river, and are aptly named the riverview apartments. The revolutionary concept completely changes the way of thinking of river front residential design. It is unnecessary to walk in order to view the river, as the view is present throughout the whole development. The apartment design must provide details of those relationships between the apartment types and landscape resources, public spaces and private spaces, window operation and position of the balcony.

Site area: 21.7 ha.
Gross floor area of Phase 2 : 165,000 m²

哈尔滨松江新城
Songjiang New City, Harbin

中国,哈尔滨 Harbin, China
业主:鲁商置业股份有限公司
Client: Lushang Property Co.,Ltd.
设计时间 Design: 2010
建成时间 Completion: 2015

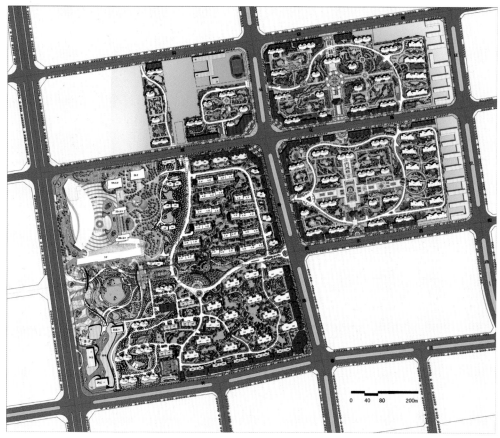

总平面图　Site Plan

用地位于哈尔滨城市中心南部，南岗区科研街、学府路、伊春路、征仪路段，为黑龙江省农科院旧址。该地区是哈尔滨市高校聚集地之一，具有浓郁的科技文化氛围，包含居住、商业、酒店、办公、公寓等业态，未来将建设成为功能完善、配套设施齐全、布局合理、特色突出的哈尔滨南部地区大型居住商业综合体。	The site is located in Xuefu road, Nangang district, south of Harbin, the former site of Heilongjiang Academy of Agricultural Sciences. The area is surrounded by colleges and universities with rich science and academic atmosphere. The project aim to be a large scale complex in south Harbin which combines residence, commercial building, hotel and office tower.
建筑设计整体定位为新欧洲主义城邦生活，体现以居住为主，综合开发的思想，以提高建筑品质，丰富空间形式，创造良好的城市形象与居住环境。设计中尽量扩大景观面，端头户型单独处理，强化区块的环境资源优势。设计还增加了每个单元的视觉通透性，尽量减少南侧建筑对北侧建筑的遮挡。	The design promotes European lifestyle with high quality building construction and diverse spatial design to create pleasant living environment. The design maximize view with special layout for the end corner apartment. Nevertheless, the design utilizes the environmental resources, minimizing the sunshade from the southern building.

总用地面积：	4.0 ha.
地上总建筑面积：	966 124 m²
容积率：	2.43

Site area:	4.0 ha.
Above ground G.F.A.:	966,124 m²
F.A.R:	2.43

青岛南山果岭艺墅
Nanshan Golf Villas, Qingdao

中国，青岛 Qingdao, China
业主：青岛南山集团长基置业有限公司
Client: Qingdao Changji Co.,Ltd.
设计时间 Design: 2010
建成时间 Completion: 2013

项目位于青岛即墨，紧邻青岛的第二大海岸鳌山湾，占地10 Km²的'海'是一座涵盖了国际博览中心、超五星级酒店、国际企业会馆、游艇俱乐部、温泉疗养中心、滨海高尔夫、写字楼与高档居住区等多种物业形态的综合型滨海度假新城。

以南北向主干道为主线，通过东西向支路连接起一个个组团，形成简洁高效的总体格局，其中主路西侧以欣赏内侧海景为主，而主路东侧则是欣赏高尔夫绿地。主路的尽端也就是基地的尽端既能享受海景，又能欣赏高尔夫绿地，自然而然成为整个项目的楼王。

The Aoshan bay, a 100,000 m² "Linghai", is a waterfront resort complex closely located from Qingdao's second biggest bay. The mixed-use facilities include a national exhibition center, a five-star hotel, an enterprise clubhouse, a yacht club, a spa treatment, a waterfront golf course, office buildings, and upscale residential areas.

The main avenue is oriented in the north-south axis while the secondary streets are oriented in the west-east axis. Each group of buildings is thoroughly connected, forming a simple and efficient distribution throughout the whole area. The west wing of the complex has a stunning view of the enclosed sea inside the complex while the east part has direct access to the vast golf course. This arrangement provides a fantastic landscape to the residents and visitors, wherever they are in the complex.

总用地面积：	1.4 ha.
总建筑面积：	42 000 m²
容积率：	0.323
总户数：	107

Site area:	1.4 ha.
Gross floor area:	42,000 m²
F.A.R:	0.323
Total amount of units:	107

INTERIOR DESIGN

室内设计

无锡千禧大酒店室内设计
Interior Design of the Millennium Hotel, Wuxi

中国，无锡 Wuxi, China
业主：无锡鑫畅置业有限公司
Client: Wuxi Xinchang Real Estate Development Co.,Ltd.
设计时间 Design: 2005
建成时间 Completion: 2009

KFS负责整个酒店的建筑设计和室内设计。

千禧酒店的室内整体设计风格延续了酒店建筑的日式风格，真正做到了室内设计是建筑设计的延续和发展。

酒店地上共有22层，地下一层，3层至22层是酒店客房。地下一层、一层、二层是公共区域。其中地下一层布置了全日餐厅、超市、地下停车场等，一层有大堂、大堂休息吧、日式餐厅、游泳池、浴场和健身房，二层是宴会厅、会议室、中餐厅和水疗等功能空间。

大堂的室内空间采用了半开放式设计，充分使用了将室外景观瀑布局部下沉的手法来分隔空间，既保证了休息区客人的私密性和相对独立性，又不影响客人观看景观。设计将瀑布美景引入室内，作为大堂休息区的背景墙，使客人从进入大堂的那一刻起就能第一眼观看到景观瀑布。由于大堂休息区域低于室内地面，所以在休息区的客人一点不会遮挡进入大堂客人的视线，充分实现了室内外景观共享的设计理念。大堂顶部设计采用弧顶来代替平顶，在弧顶中间加有日式图案丰富弧顶造型，弧顶的图案又与背景墙的图案形成呼应，既统一又和谐。

24小时餐厅在地下一层，靠近室外景观瀑布的一边采用玻璃隔断，正如大堂一样，将室外的景观引入室内。站在这里根本感觉不到这是地下一层，而更像是站在了地上一层或二层的感觉。而伸出去的平台设计增加了人们亲临瀑布的真实感，做到了人与大自然最有机的融合。在午后的斜阳中品着美酒，吃着牛排，看着阳光斜照，欣赏着瀑布下落泛起的小浪花，听着优美动听的音乐，夕阳西下给繁忙的都市人带来的情调。

日式餐厅的室内设计则完全按照日本餐厅的布局和形式进行设计，能够充分唤起在无锡游走的日本旅客对故乡的思念，使进入餐厅的客人刹那间有一种归属感。

总统套房室内设计为了实现富丽堂皇的感觉，在设计上采用了欧式风格进行装饰，无论是复杂的装饰线条及其顶上金箔的运用，还是真皮家具的使用，无处不体现了人们对奢华氛围的奢望和追求。

客房的室内采用了日式窗户的分隔方法，在白色乳胶漆的墙面上加了木框线条进行分隔，日式风格十足。房间的家具也采用了日式家具。整体设计做到了风格的统一。

KFS provides full service designing of interior and exterior architecture for the hotel. The style of the interior of the Millennium Hotel follows the Japanese exterior style. The interior design retain the continuity of the architectural exterior.

The hotel has 22 floors and a lower level. The hotel rooms are from the third floor up. The lower level, ground floor and first floor are public spaces. The lower level accommodates a Japanese restaurant, supermarkets and parking. The lobby, waiting room, Japanese restaurant, swimming pool, bathhouse and gym are located at the ground floor. A banquet hall, conference room, Chinese restaurant and spa are located on the first floor.

The interior of the lobby is a semi-open design that realizes the partition of intimate and public space by using the exterior landscape. The design introduces the artificial waterfall as the background for the lobby. A sunken floor for the lobby makes the waterfall a prominent element and visually connects the main entrance and the landscape. The concept of landscape sharing is used throughout the public interiors. An arc-formed roof replaced the commonly used flat design. KFS implemented the traditional Japanese pattern to enrich the shape of the ceiling. The patterns of the roof respond to the decorative patterns on the walls in a harmonious resonance.

A 24-hour restaurant is located at the lower level. A part of the restaurant that is adjacent to the outer landscape is designed in glass. The design concept of introducing the landscape into the interior is similar to the lobby design. Dining in the restaurant, although below ground, the same feeling and pleasure as the ground floor or even above is maintained. The extended exterior terrace offers a great intimacy. The waterfall, integrating man and nature into a dynamic harmony, painting the image that savours a tender steak, accompanied by a good glass of wine amidst the subtle and soothing sounds of water blended in with the soft music.

The Japanese restaurant is designed in accordance with the layout and interior of a traditional Japanese restaurant to create a nostalgic atmosphere for the Japanese guests in Wuxi seeking a atmosphere similar to their eating places in Japan. The achievement of the feeling of grandeur in the Presidential Suite a luxury interior design in European style was chosen. The decorative elements of golden leafs, delicate lines and the leather furniture, all contribute to an atmosphere of luxury.

The interiors of the rooms are designed according to the Japanese style of painting the walls white with wooden frames to complement the window opening and completed with Japanese-styled furniture.

上海达安圣芭芭花园河谷3号——90花园别墅室内设计
Interior Design of St.Babara Valley No.3—90 Villa, Shanghai

中国，上海 Shanghai, China
业主：上海达安泰豪置业有限公司
Client: Shanghai Da'an Taihao Real Estate Co.,Ltd.
设计时间 Design: 2009
建成时间 Completion: 2009

本项目的室内设计延续了建筑的空间感,并且进一步强化了空间的流动感和交流度,在感官上扩大了室内空间。比如,地下室空间对庭院的扩大和延续,卧房中睡眠区和盆浴区的统一和连通,卧室空间与室外景区的沟通和交流。

河谷3号样板间共有五种房型,其室内设计风格基本上也都采用了地中海地区国家的风格进行设计,有西班牙风格、现代欧式风格、现代风格等。由于建筑设计时对建筑的室内空间进行了思考,五种房型的室内空间基本保持建筑设计的分隔空间,只是在立面和地面的装饰材料上有所区分。

设计在河谷3号每种房型地下室都开有天井,这样就主动把室外阳光引进室内,彻底颠覆了人们对地下空间的通常印象和认识,再在天井下方或放绿色植物或摆放休闲长椅。有长椅、有植物、有阳光,是室外还是室内?是地上还是地下?完全由主人的感受而定。

在主卧室设计上,设计大胆地采用了卧室与卫生间无隔断或采用透明玻璃隔断的方式,或在床的正上方开天窗或在浴缸的上方开天窗,充分利用建筑设计过程中遗留的优势进行重组并合理分配。白天能够把阳光引入室内,使天窗下方的空间特别明亮,视觉冲击力强。晚上更是能够看月亮,数繁星。这种大胆而有创意、温情而浪漫的设计,不仅节省了空间,创造了价值,而且增加了神秘感和趣味性,给人留下了深刻的印象。

马赛克在整个房型也有大量的运用或以水的颜色——蓝色进行铺设,或以暖色橘红色进行铺设,或以浅色乳白色加灰色进行铺设。由于马赛克单位面积小,所以比较灵活,能够更加完美地演绎设计风格,给室内空间带来丰富多彩的视觉效果。由于所有房型的整体面积都比较小,为了节约空间,留给楼梯的空间非常狭小,为了满足人的视觉的舒适性,消除狭小空间给人的带来的压抑感,设计采用透明玻璃隔断来代替普通的墙体隔断,开阔了人的视野,改善了人的感观舒适性。

The interior design retains the same architectural special spatial feeling. The design emphasizes the continuity of the internal space to amplify the effect of special spatial perception. The open lower level space is an extension of the garden. The extension of the bedroom space into the bathroom space, and linkage of the bedroom space with the exterior landscape.

Valley No.3 has a total of five different unit models. The interior design of the five models is Mediterranean in style and responds to the outer facades. The design of the interior retains special spatial separation. In accordance with the architectural design the decorative facade and materials of the ground floor differ from the upper levels.

All villas of Valley No.3 have an open courtyard at the lower level to introduce sun light into the building. This design consideration completely transforms the stereotype of the underground space having less desirable conditions. The design of the yard as an extension of the room, through the placing of vegetation and furniture, softens the division between the indoor and exterior living space.

The bathroom is integrated in the master bedroom. In some rooms KFS makes use of transparent glass to set the bathroom apart from the bedroom creating an illusion of a complete separation. Some sky windows have been designed above the bed and the bathtub to make these parts of the room particularly bright. In the evening the sky windows stimulate the strong visual impact derived from seeing the moon and the stars from the interior. This bold, creative, loving and romantic design saves space, and creates value and an increased sense of mystery.

The villas are extensively decorated with different colored mosaics in a background of light gray and white mosaic tile. The experience and enjoyment of visual stimulus is everywhere. Color has been used extensively to denote function and use. The blue mosaic recalls the presence of water. The warm orange colored mosaic announces an area intended to relax. The rooms are compact in size and KFS made extensive use of glass walls to enhance and visually enlarge the spatial effect.

KFS国际建筑师事务所上海办公楼室内设计
Interior Design of KFS Office, Shanghai

中国，上海 Shanghai, China
业主：加拿大KFS国际建筑师事务所
Client: KFS STONE DESIGN INTERNATIONAL INC, CANADA
设计时间 Design: 2009
建成时间 Completion: 2010

设计师在建筑设计时已对室内空间进行了思考和规划，因此在室内设计时充分尊重和延续了建筑的空间感，并且进一步强化了空间的流动感和交流度，在感官上扩大了室内空间。

整个设计格调是简洁大方的现代办公室。设计大量运用玻璃隔断，模糊了室内和室外的空间，把室外的景观引入到室内来。

接待处一进门就可看到几只镇室之兽整齐地摆放在墙的一侧，接待台上方火红的四盏灯笼灯预示着办公室的生意红红火火，且与周围的浅色调形成鲜明的对比，凸显其个性。

每个房间门中间都是透明玻璃，玻璃背后有罗马帘，与外界沟通全靠罗马帘来控制。把罗马帘拉上去，可以很清楚地看到屋外发生的一切；把罗马帘拉下来，屋内就是一个相对私密的空间。

室内大办公室采用开放式设计，给在这里工作的人以设计的通透性，使设计的激情在这里可以得到尽情地释放。

楼与楼之间的中庭，挑空了三层，上面采用半封闭式设计，一半玻璃窗，一半实体顶面。前后门和窗都采用玻璃进行与外界的分隔，玻璃窗直通顶部，使本来有点狭小的空间一下子高耸起来，并且把前后花园的景色完全引入中庭，体现了中国建筑中庭院的概念。

The architectural design of the interior space was done through careful planning and consideration. The interior design is completely integrated the architectural sense of space. The design strengthens the continuity and movement of space to amplify the importance of spatial perception.

The theme of the office design is simple, modern and elegant. Extensive use of glass in partitioning spaces eliminates defined boundaries between interior and exterior space. The exterior landscape is transformed into an interior design feature to enhance the interior spatial quality.

Upon entering the reception area the statues-of-grounding can be seen as four mythological creatures in a row in contrast to the neutral interior wall behind. The interior lighting features for reception area are four red lanterns that indicate the conduct of brisk business and the full traditional eastern personality of the office.

The doors are transparent glass. The Roman blinds, installed behind the glass, control the relation between the interior offices and the exterior landscaped gardens. When the Roman blinds are in the open position the connection to exterior is self-evident. When the Roman blinds are down the office is transformed into an secluded intimate space.

The open-style design of large office provides a clear and pleasing environment for the staff and provides a room and space devoted to creative thinking.

The three-storey high atriums between each building have semi-enclosed roofs with half-glazed roof lighting. The rear door and door to the street are fully glazed to link the atrium with the exterior space, making the moderate-sized room to be without walls. These atriums exhibit the traditional Chinese courtyard-philosophy of integration with the front garden and the back garden.